Palgrave Advances in Luxury

Series Editors
Paurav Shukla
Southampton Business School
University of Southampton
Southampton, UK

Jaywant Singh
Kingston Business School
Kingston University
Kingston Upon Thames, UK

The field of luxury studies increasingly encompasses a variety of perspectives not just limited to marketing and brand management. In recent times, a host of novel and topical issues on luxury such as sustainability, counterfeiting, emulation and consumption trends have gained prominence which draw on the fields of entrepreneurship, sociology, psychology and operations.

Examining international trends from China, Asia, Europe, North America and the MENA region, *Palgrave Advances in Luxury* is the first series dedicated to this complex issue. Including multiple perspectives whilst being very much grounded in business, its aim is to offer an integrated picture of the management environment in which luxury operates. It explores the newer debates relating to luxury consumption such as the signals used in expressing luxury, the socially divisive nature of luxury and the socio-economic segmentation that it brings. Filling a significant gap in our knowledge of this field, the series will help readers comprehend the significant management challenges unique to this construct.

All submissions are single blind peer reviewed. For more information on our peer review process please visit our website: https://www.palgrave.com/gp/book-authors/your-career/early-career-researcher-hub/peer-review-process.

This series is indexed in Scopus.

Claudia E. Henninger
Navdeep K. Athwal
Editors

Sustainable Luxury

An International Perspective

Editors
Claudia E. Henninger
Department of Materials
The University of Manchester
Manchester, UK

Navdeep K. Athwal
School of Business
University of Leicester
Leicester, UK

ISSN 2662-1061   ISSN 2662-107X (electronic)
Palgrave Advances in Luxury
ISBN 978-3-031-06927-7   ISBN 978-3-031-06928-4 (eBook)
https://doi.org/10.1007/978-3-031-06928-4

© The Editor(s) (if applicable) and The Author(s), under exclusive licence to Springer Nature Switzerland AG 2022
This work is subject to copyright. All rights are solely and exclusively licensed by the Publisher, whether the whole or part of the material is concerned, specifically the rights of translation, reprinting, reuse of illustrations, recitation, broadcasting, reproduction on microfilms or in any other physical way, and transmission or information storage and retrieval, electronic adaptation, computer software, or by similar or dissimilar methodology now known or hereafter developed.
The use of general descriptive names, registered names, trademarks, service marks, etc. in this publication does not imply, even in the absence of a specific statement, that such names are exempt from the relevant protective laws and regulations and therefore free for general use.
The publisher, the authors, and the editors are safe to assume that the advice and information in this book are believed to be true and accurate at the date of publication. Neither the publisher nor the authors or the editors give a warranty, expressed or implied, with respect to the material contained herein or for any errors or omissions that may have been made. The publisher remains neutral with regard to jurisdictional claims in published maps and institutional affiliations.

This Palgrave Macmillan imprint is published by the registered company Springer Nature Switzerland AG.
The registered company address is: Gewerbestrasse 11, 6330 Cham, Switzerland

Navdeep—I dedicate this co-edited book to my darling husband and daughter.
Claudia—I dedicate this co-edited book to Jutta and Hartmut.

Preface

Navdeep and Claudia met as early career researchers at University of Sheffield Management School in 2015. They immediately connected over their mutual passion to explore the world and love of all things fashion. They had the idea of publishing a co-edited book on sustainable luxury many years ago, although it took them a few years to get round to writing the book proposal as life got in the way. They thank the authors for their valuable contributions to this book, for persevering during the COVID-19 pandemic and for researching in a field that they benefit will raise more awareness of the ecological impact of the luxury sector.

Manchester, UK Claudia E. Henninger
Leicester, UK Navdeep K. Athwal

Contents

1 Introduction 1
Navdeep K. Athwal and Claudia E. Henninger

2 The Hidden Value of Second-Hand Luxury: Exploring the Levels of Second-Hand Integration as Part of a Luxury Brand's Strategy 13
Linda Lisa Maria Turunen and Claudia E. Henninger

3 Sustainable Eco-luxury in the Scandinavian Context 35
Kirsi Niinimäki

4 Sustainable Luxury: A Framework for Meaning Through Value Congruence 59
Stephanie Y. Volcon, Marian Makkar, Diane M. Martin, and Francis Farrelly

5 Sustainability, Saudi Arabia and Luxury Fashion Context: An Oxymoron or a New Way? 81
Sarah Ibrahim Alosaimi

6   Towards Circular Luxury Entrepreneurship: A Saudi
    Female Entrepreneur Perspective   101
    *Rana Alblowi, Claudia E. Henninger, Rachel Parker-Strak,
    and Marta Blazquez*

7   Sustainable Supply Chain Process of the Luxury *Kente*
    Textile: Introducing Heritage into the Sustainability
    Framework   129
    Sharon Nunoo

8   Canadian Ethical Diamonds and Identity Obsession:
    How Consumers of Ethical Jewelry in Italy Understand
    Traceability   153
    Linda Armano and Annamma Joy

9   Sustainability Claims in the Luxury Beauty Industry: An
    Exploratory Study of Consumers' Perceptions and
    Behaviour   173
    Panayiota J. Alevizou

10 What Do You Think? Investigating How Consumers
    Perceive Luxury Fashion Brand's Eco-labelling Strategy   197
    *Shuchan Luo, Aurelie Le Normand, Marta Blazquez, and
    Claudia E. Henninger*

11 'Take a Stand': The Importance of Social Sustainability
    and Its Effect on Generation Z Consumption of Luxury
    Fashion Brands   219
    Helen McCormick and Pratibha Ram

12 Chinese Consumer Attitudes Towards Second-Hand Luxury Fashion and How Social Media eWoM Affects Decision-Making 241
Rosy Boardman, Yuping Zhou, and Yunshi Guo

13 The Rise of Virtual Representation of Fashion in Marketing Practices: How It Can Encourage Sustainable Luxury Fashion Consumption 271
Shuang Zhou, Eunsoo Baek, and Juyeun Jang

Index 293

Notes on Contributors

Rana Alblowi is a faculty member at Prince Sattam Bin Abdulaziz University, KSA, where she has undertaken many roles within the university. Rana holds a master's degree in Apparel, Merchandising, Design and Textiles at Washington State University, USA, and is a PhD researcher in Fashion Marketing Management at The University of Manchester, UK. Her thesis and research are into entrepreneurship and fashion sustainability. She has presented her research at numerous conferences worldwide. In addition to her research activities, she holds lectures and leads workshops on entrepreneurship, entrepreneurs' creativity and innovation, and sustainable fashion practices.

Panayiota J. Alevizou is Lecturer in Marketing at the Sheffield University Management School, where she teaches Marketing. Her research focuses on sustainability labelling, eco-labelling and sustainability marketing in the context of FMCGs and fashion. Panayiota has presented her work in various national and international conferences, and her articles have appeared in journals such as *Journal of Marketing Management*, *European Journal of Marketing*, *Journal of Fashion Marketing and Management*, *Journal of Business Research* and *Journal of Consumer Behavior*. She is co-chair of the Sustainability Special Interest Group (SIG) of the Academy of Marketing.

Sarah Ibrahim Alosaimi is Lecturer in Fashion Management and Marketing at Princess Nourah Bint Abdul Rahman University. Sarah holds a BA in Fashion and Textile Design, Princess Nourah Bint Abdul Rahman University, and a master's degree from the University of Southampton, where she specialised in Fashion Marketing and Branding. She was awarded a PhD in Fashion Management and Marketing from The University of Manchester, United Kingdom. Her research focuses on social media marketing, sustainability in the fashion industry, consumer behaviour, and consumption from both luxury and cultural perspective. She has been invited to multiple lectures on several occasions at the University of Southampton, and she also regularly attends academic conferences to present her work at international conferences.

Linda Armano is Marie Curie Fellow at Ca' Foscari University of Venice in the Department of Management, and at the University of British Columbia (UBC) in the Faculty of Management. Her investigation looks at the mines in the Northwest Territories of Canada and in Italian jewelry stores that sell Canadian diamonds in Italy. Her recent publications are *The Alpine Mining Culture. Self-Representation of Professional Miners* (Aracne Editrice), "Doing Well While Doing Good. How the Hybrid Business Model Promotes Sustainability in the Fashion Industry" (co-authors Annamma Joy, Camilo Peña) in the *Journal of Business Anthropology*; Encoding Values and Practices in Ethical Jewelry Purchasing: A Case History of Ethical Luxury Consumption (co-author Annamma Joy), in Coste-Maniere, Miguel Angel Gardetti (eds), *Sustainable Luxury and Jewellery* (Springer Nature).

Navdeep K. Athwal has held academic positions in the UK and Hong Kong, where she researched the interplay between sustainability and luxury. Her work has been published in a range of academic journals such as *Journal of Marketing Management, Information Technology and People*, and *International Journal of Management Reviews*.

Eunsoo Baek is an assistant professor at The Hong Kong Polytechnic University. Baek's primary research interests center on understanding consumers in fast-changing fashion and retail environments based on consumer psychology discipline. Her ongoing and completed research projects aim to understand how retail elements affect shoppers. Her recent articles demonstrate her keen interest in interactive technologies (AR/VR) in retail contexts, including topics of a virtual tour to a retail store and AR try-on experience.

Marta Blazquez is Senior Lecturer in Fashion Marketing. Her research interest includes retail marketing, store futures, technology and digital sustainability. Marta's research has been published in the *Journal of Business Research, Computers in Human Behaviour* and *International Journal of Electronic Commerce*, among others, and she has edited several books on sustainable fashion and social commerce. She holds a BA in Advertising and PR and has developed a career in the advertising and marketing industry for more than ten years. She is further a co-chair of the AoM's SIG Sustainability.

Rosy Boardman is Senior Lecturer in Fashion Business at the Department of Materials, The University of Manchester. Rosy's research focuses on digital strategy and innovation in the fashion retail industry. In particular, her research specialises in ecommerce, digital marketing and social media marketing, using eye-tracking technology and qualitative research methods. Her interest is in exploring fashion retail's current and future developments, focusing on the digital economy and consumer behaviour as well as how technology can be used to solve issues related to both social and environmental sustainability. Rosy has published several peer-reviewed academic papers in world-leading journals as well as books, including *Social Commerce: Consumer Behaviour in Online Environments* (Springer, 2019) and *Fashion Buying and Merchandising in a Digital Society* (2020).

Francis Farrelly is Professor of Marketing at RMIT University. His work has appeared in journals such as *the Journal of Marketing, the Journal of Consumer Research, Annals of Tourism Research*, and *the Harvard Business Review*. His research interests include consumer insights, consumer culture, marketing strategy, and tourism.

Yunshi Guo completed an MSc in International Fashion Retailing at the Department of Materials, The University of Manchester, and now works as a practitioner in the field.

Claudia E. Henninger is Senior Lecturer in Fashion Marketing Management, with a research interest in sustainability, the circular economy, and more specifically collaborative consumption, in the context of the fashion industry. Her work has been published in *the European Journal of Marketing*, *the Journal of Fashion Marketing & Management*, and *the International Journal of Management Review*, and she has edited a variety of books on sustainable fashion, including *Sustainability in Fashion – A Cradle to Upcycle Approach*. Claudia is further the Chair of the Academy of Marketing's SIG Sustainability and an Executive Member of the Sustainable Fashion Consumption Network.

Juyeun Jang is a research assistant professor at The Hong Kong Polytechnic University. Jang's primary research focuses on immersive retailing using advanced technology, such as VR, and its impact on consumers' information processing and behavioral response for developing creative retail environments and business strategies. She also has carried out innovative and practical research projects to investigate the impact of on/offline store environments on consumers' cognitive/affective and psychophysiological responses applying eye-tracking, heart rate, muscle activities, and others.

Annamma Joy is Professor of Marketing at the Faculty of Management, UBC. Her research spans the domains of art, fashion and fine wines. She has won several awards for her research, the most recent being the Louis Vuitton and Singapore management university award (second place) for best paper at the Luxury brand conference in 2018. Her work has been published in *the Journal of Consumer Research*; *Journal of Consumer Psychology*; *Journal of Business Research*, *Consumption, Markets and Culture*; *Journal of Consumer Culture*; *Cornell Hospitality Quarterly*; *Tourism Recreation Research*; *Journal of Wine Research and Arts and the Market*. She has several chapters in handbooks on consumption and marketing. She is editing three books on the related topics of artification, sustainability and digitalization in art, fashion and wine.

Aurelie Le Normand is Lecturer in Fashion CAD at the School of Materials, The University of Manchester, and teaches on a variety of units across the BSc Fashion Business programs relating to buying & merchandising, fashion CAD, and fashion communication. Aurelie has an industry background and worked as a designer for well-known organizations in the UK. Aurelie's design work is recognized for challenging the perception of fashion as a means of design innovation.

Shuchan Luo is a PhD researcher, with a research interest in sustainability in the luxury fashion industry, sustainability communication, and sustainable material innovations, in the context of the fashion industry. She has presented her research in the Academy of Marketing Conference. Her work has been published in *Journal of Design Business & Society*.

Marian Makkar is Lecturer in Marketing at the College of Business, RMIT University, Melbourne, Australia. Her research interests include consumer culture, market shaping, the sharing economy, and tourism experiences. Her research has appeared in international journals such as *European Journal of Marketing*, *Journal of Business Research*, *Current Issues in Tourism*, *Journal of Fashion Marketing & Management*, *Journal of Retailing and Consumer Services*, and *Marketing Intelligence & Planning*.

Diane M. Martin is Professor of Marketing and the leader of the Consumer Culture Insights Group at RMIT University in Melbourne, Australia. Her work has appeared in numerous top journals and is the co-author of the book *Sustainable Marketing*. Her award-winning research employs ethnographic methods in relationships between consumers, communities and culture. She also conducts research which identified marketing as a solution to the problem of sustainability. Martin is American Leadership Forum Senior Fellow.

Helen McCormick is Reader in Fashion Marketing within the Manchester Fashion Institute. As a subject specialist in fashion marketing Helen conducts research in the areas of digital strategy and innovation in the retail industry, focusing on consumption behaviour, marketing communications and customer engagement and sustainable business practices in fashion retail and responsible consumption behaviour management. Her research interests follow contemporary issues considering how they affect consumer and business practices. Helen has an established track record of publishing in internationally recognised journals as well as attending marketing, management, branding and technology conferences. She has been invited as a keynote speaker at a number of events concerning her research utilising eye-tracking technology and the changing retail environment. She has significant experience in leadership roles in higher education in teaching and learning and line management providing leadership direction and strategic operational management with regard to learning, teaching and the student experience.

Kirsi Niinimäki is Associate Professor of Design, especially in Fashion research in the Department of Design at Aalto University, Finland. Her research has focused on holistic understanding of sustainable fashion and textile fields and connections between design, manufacturing, business models and consumption. She addresses the subject of sustainability from a variety of perspectives, and her scientific results create a cohesive collection of new knowledge in this field. Her research group, the Fashion/Textile Futures, http://ftfutures.aalto.fi, is involved in several significant research projects, which integrates closed loop, bio-economy and circular economy approaches in fashion and textile systems and extends the understanding of strategic sustainable design. Niinimäki has published widely in top scientific journals, and she is a leading scholar in the field of sustainable fashion.

Sharon Nunoo is a PhD researcher at The University of Manchester with interest in cultural heritage, communities and corporate marketing. Her research was present at the Academy of Marketing Conference in 2019, which is based on exploring corporate marketing strategies within cultural heritage communities. Her work has been in the edited book on *Handloom Sustainability and Culture* by Gardetti, M.A. and Muthu, S.S.

Rachel Parker-Strak is Senior Lecturer in Fashion Buying and Merchandising at The University of Manchester. Her teaching and research focus is on fashion product development and buying and merchandising. Prior to joining academia, Rachel's industry career was in product development positions for a UK-based womenswear brand.

Pratibha Ram is a lecturer at The University of Manchester. She has been teaching bachelor and master programs for almost ten years. Prior to joining The University of Manchester, she was a senior lecturer in Singapore and a lecturer in Hong Kong. She started her career in marketing and gained several years of industry experience, where she held various positions in business development, marketing, advertising, marketing, and academic research. Her industry experience has shaped her research interests which are primarily consumer research and branding. She is particularly interested in themes that relate to the social and cultural impact of branding and consumption.

Linda Lisa Maria Turunen is a post-doctoral scholar at Aalto University School of Business. Her research interests and specialization lie in brand management and consumer behavior, particularly in the context of luxury fashion brands and perceived luxuriousness, luxury second-hand and vintage, fashion resale and sustainability. Her several papers have appeared in major international journals, such as *Journal of Business Research*, *Journal of Cleaner Production*, *Journal of Product and Brand Management*, and *International Journal of Consumer Studies*, and she has contributed to various book chapters as well as popularized her research through the national press. She holds a PhD from University of Vaasa and has been granted the title of docent, Fashion and Marketing research, in University of Turku, Finland.

Stephanie Y. Volcon is a PhD candidate in Marketing at RMIT University, Melbourne, Australia. Her research focuses on the cultural analysis of sustainability, luxury, and status products within consumer culture theory. She has worked in various corporate marketing roles including strategic brand planner, brand manager, and digital account manager of beauty and luxury brands.

Shuang Zhou is Research Assistant Professor in Fashion Business at The Hong Kong Polytechnic University. She obtained her PhD in Fashion Marketing at The University of Manchester in 2019. Her research interests focus on influencer marketing, marketing communications, technological innovations in marketing and retail, social media marketing, consumer behaviour, and luxury consumption. She has presented her research at several national and international academic conferences and published journal articles related to influencer marketing.

Yuping Zhou completed an MSc in International Fashion Retailing at the Department of Materials, The University of Manchester, and now works as a practitioner in the field.

List of Figures

Fig. 2.1	Three main strategies of second-hand integration for luxury brands	29
Fig. 5.1	Paradoxical framework for sustainability and luxury within the Saudi context	91
Fig. 7.1	Simplified supply chain process of *Kente* strips (Fening, 2006; Lartey, 2014; Amissah & Afram, 2018; Thirumurugan & Nevetha, 2019)	136
Fig. 8.1	Conveyance of the concept of ethical diamond traceability to Italian ethical consumers	164
Fig. 11.1	Conceptual framework social sustainability/activism	228
Fig. 13.1	Implications of virtual representation of fashion (VRF) for encouraging sustainable luxury fashion consumption (SLFC)	283

List of Tables

Table 1.1	Characteristics of luxury brands	2
Table 2.1	Three strategies of second-hand integration in luxury resale	22
Table 3.1	Respondents' age	41
Table 3.2	Respondents' income levels	42
Table 3.3	Sources of luxury and frequency of searching for it	42
Table 3.4	Positive image of colour and fibre sources (natural or synthetic)	42
Table 3.5	What elements affect consumers' experience of luxury (in clothing)?	43
Table 3.6	Finnish consumers understand luxury through three different categories	44
Table 4.1	Common values intrinsic to both luxury and sustainability	65
Table 6.1	Summary of participants	112
Table 9.1	Participant profiles and purchases	182
Table 10.1	Overview of participants	204
Table 10.2	Consumers' perspective about the influence of eco-labels (objective vs. subjective)	207
Table 12.1	Interview information	265
Table 13.1	VRF stimuli utilised in luxury fashion marketing touchpoints	275

1

Introduction

Navdeep K. Athwal and Claudia E. Henninger

1.1 What Is Sustainable Luxury?

The purpose of this introductory chapter is to provide some clarification on definitions and the context of this book, before providing an overview of the individual contributions incorporated in this edited collection.

Looking at the title of this edited volume, it contains chapters surrounding sustainability combined with sustainability in the fashion industry. Focusing on luxury first, it originates from the Latin word 'luxuria' and literally translates to *extravagance* or *excess*. Turunen (2018) highlights that historically speaking luxury can have both positive and negative connotations. Positive in that luxury is seen as something that may be aspirational and thus could encourage individuals to not only improve themselves but also society. Contrary to this, it has been

N. K. Athwal
School of Business, University of Leicester, Leicester, UK

C. E. Henninger (✉)
Department of Materials, The University of Manchester, Manchester, UK
e-mail: Claudia.Henninger@manchester.ac.uk

negatively perceived and described as the enemy of virtue (Kapferer & Bastien, 2009).

Independent of the associations one may have, luxury items are commonly those that are 'additional' and thus may no longer fulfil a need but rather a want (Li et al., 2012). Yet, herein lies the challenge. Our society has changed dramatically over the centuries and what was previously seen as 'luxury' may no longer be the same today. Is luxury something that is unique and imported from different cultures, or is it an item that comes with a high price take and (ideally) high quality, or is it an experience? (Kapferer & Bastien, 2009; Chevalier & Mazzalovo, 2012; Athwal et al., 2019). The fact is, luxury can be anything depending on the individual and their interpretation, which makes it an exciting topic to investigate (Ryding et al., 2018; Turunen, 2018; Athwal et al., 2019).

To provide a clear understanding however, within this edited volume, luxury and more specifically luxury brands are understood as those that have some or all of the characteristics outlined in Table 1.1.

Now that we have outlined the range of meanings associated with luxury, it is essential to discuss sustainability as it has also been described as a fuzzy concept that can be intuitively understood (Henninger et al., 2016, 2017; Niinimäki, 2018). Sustainability is traditionally portrayed as an overlapping Venn diagram, whereby the intersection of social, environmental, and economic aspects is what is commonly referred to as 'sustainability' (e.g., Elkington, 2004). More recently, we see sustainability discussed through Earth Logic (Fletcher & Tham, 2019), which centres on the harmony between people and planet thereby setting out an action plan of six principles: (1) reduce growth, (2) refocus on the local, (3) develop new centres for fashion, (4) learn new competences, (5) showcase

Table 1.1 Characteristics of luxury brands

Characteristics	References
Premium price	Phau and Prendergast (2000), Okonkwo (2007)
High quality	Ryding et al. (2018), Athwal et al. (2019)
Rich history and/or heritage	Turunen (2018)
Scarce, exclusive, unique	Okonkwo (2007), Keller (2009)
Often from designers	Phau and Prendergast (2000), Okonkwo (2007)

a different approach of communication for fashion, and (6) create new ways of organising fashion through governance.

To reiterate, it becomes apparent that sustainability remains an intuitively understood concept that has a variety of definitions and interpretations (e.g., Henninger et al., 2016, 2017; Niinimäki, 2018), some of which are reflected within the contributions of this book. As alluded to, sustainability is however not the only focus of this edited volume, but rather it is contextualised within the luxury fashion sector, where it has become a pervasive issue gaining traction with brand managers, scholars, policy-makers, the media and academia (Athwal et al., 2019; Davies et al., 2020; Wells et al., 2021). The luxury industry is a key market segment that accounted for US $117,152 million in 2022 (Statista, 2022) and predicted to grow 4.27% annually in the coming years, making it interesting to investigate further. Caïs (2021) further highlights that "the ironic paradox of sustainable luxury brands remains prevalent," which has also received increased attention within the literature (e.g., Athwal et al., 2019; Brydges et al., 2020; Luo et al., 2021).

What do we mean by this paradox and why is this of interest? Especially since the COVID-19 pandemic we have seen a significant focus on sustainability, with various outlets either posing questions, such as "Could the Covid pandemic make fashion more sustainable?" (Marriott, 2020), "Will COVID-19 accelerate the transition to a sustainable fashion industry?" (Ricchetti & De Palma, 2020), or more explicitly highlighting "How Covid turned the spotlight on sustainability" (Fish, 2020). Either way, sustainability has a spotlight position and is increasingly a focal point for consumers, who have been observed to change their behaviour and increasingly demand more sustainability measures (Granskog et al., 2020; Iran et al., 2022).

More broadly, the global fashion industry is one of the most polluting, and is often criticised for unsustainable practices, including, but not limited to, waste creation, use of chemicals, treatment of workers along the supply chain, and burning of unsold goods (Blazquez et al., 2020; Brydges & Hanlon, 2020; Dickson & Warren, 2020). Yet, "luxury brands have an advantage over fast fashion companies—their products are purchased for longevity, rather than being temporary and disposable like their inexpensive counterparts" (Davis, 2020). Whilst disposability is less of an issue in the luxury fashion industry, and thus, sustainability may have, in the past, not been as much of an issue, its raw materials are—seeing as luxury goods are often made from animal-based raw materials or precious metals (e.g., Henninger et al., 2022). Thus, it may not be surprising that sustainability is increasingly priorities now in the luxury fashion industry (e.g., Athwal et al., 2019). We increasingly see luxury fashion brands advertising vegan alternatives within their fashion lines (e.g., Stella McCartney) or using different production techniques, such as upcycled vintage collections (e.g., Miu Miu) (Bala, 2021; Henninger et al., 2022).

The purpose of this edited book is to examine the state of sustainable luxury research in marketing and consumer behaviour across cultures and geographical regions. The luxury market spans sectors and innately international, while luxury consumers are characteristically a nation's most elite and wealthiest individuals. The reach of luxury markets brings the expectation that research exploring sustainable luxury will reflect multiple national and international markets. This book reflects luxury consumption as it addresses the significant scholarly omission of research by exploring culture and sustainable luxury.

When editing this book, we had an academic audience in mind, particularly as we had seen an emergence in scholarly activity on topic from 2003 (see Athwal et al., 2019). We believe that this book will be a key resource to better understand sustainable luxury and the fashion industry. We welcome scholars and educators to use this book as either a supplementary or core reading as it will provide students with a wealth of information on sustainable luxury and the fashion industry within different country contexts. There is no doubt that this book will also be beneficial to practitioners due to its key business insights. During the peer-review process, the authors were asked to consider the managerial implications from their research findings.

1.2 About This Book

This book brings together a collection of scholars spanning many disciplines such as marketing, management, textiles, fashion, economics, and digital media, as well as ranging in the geographical scope to reflect the book's international perspective. The diversity of chapters is also reflected in the combination of empirical and conceptual works, while the authors themselves range from emerging scholars to internationally leading academics.

Whilst in the past luxury may have been associated predominantly with new, novel, and unique items, what is luxury has shifted, as was outlined in the introduction. It may not be surprising that we see 'new' ways of how luxury fashion retailers/brands are engaging with the ever-increasing second-hand luxury market (Turunen & Leipämaa-Leskinen, 2015; Turunen et al., 2020; Beauloye, 2021; Rooks, 2021). We also see that our understanding of luxury fashion has changed and may have now a different meaning that it originally had.

Chapter 2 addresses the former aspect, in that it provides more detailed insights into how industry players are dealing with a shift in consumer behaviour, which sees the second-hand movement on the rise. Within Chap. 2, **Linda Lisa Maria Turunen** (Aalto University, Finland) and **Claudia E. Henninger** (The University of Manchester, UK) explore '*The Hidden Value of Second-Hand Luxury: Exploring the Levels of Second-Hand Integration as Part of a Luxury Brand's Strategy.*' The chapter takes on a brand perspective, thereby outlining different options currently available on how they (luxury fashion retailers) can engage with this lucrative market segment, which includes items ranging from handbags to watches, and has an estimated worth of €33 billion/$37.2 billion (Granskog et al., 2020). Data collection for this chapter is based on secondary resources, which have been carefully analysed and critically evaluated. As a result, the authors provide three strategies of second-hand integration in luxury resale and what the implications may be if these are to be executed.

Chapter 3 addresses the latter aspect in that it critically examines the meaning of luxury and illustrates that there is indeed a shift in what we as consumers see to be luxury. **Kirsi Niinimäki** (Aalto University, Finland) investigates '*Sustainable Eco-luxury in the Scandinavian Context.*' The chapter is based on the hypothesis that luxury in the Scandinavian context is connected to sustainable values, eco-appreciation, natural-based alternatives, and good 'everyday' design. Conducting both a consumer study and a company analysis, findings outline a key finding that may be more surprising: consumers expected luxury garments to be more sustainable, have a more transparent production process, and that the companies behind luxury products are more responsible than the average fashion company. Key theoretical contributions and managerial implications conclude this chapter.

Staying within the remit of culture and coming back to a paradox highlighted in the introduction, Chap. 4, authored by **Stephanie Volcon, Marian Makkar, Diane M. Martin, and Francis Farrelly** (RMIT University, Australia), focuses on '*Sustainable Luxury: A Framework for Meaning Through Value Congruence.*' Through a case study approach the chapter hones in on aspects of language, symbols, and artefacts used within a sustainable luxury brand and outlines how the cues can be utilised to attract sustainably conscious consumers. The chapter takes on a conceptual approach and concludes with areas of future research.

Chapter 5 follows on nicely from the conceptual layout in Chap. 4, in that it investigates '*Sustainability, Saudi Arabia and Luxury Fashion Context: An Oxymoron or a New Way.*' **Sarah Ibrahim Alosaimi** (Princess Nourah Bint Abdul Rahman University, Saudi Arabia) provides a data-driven chapter, outlining issues surrounding sustainability in a country that more recently made quite dramatic changes in fostering sustainability. The chapter highlights that there is a challenge not only with sustainability and luxury but also with how luxury is interpreted within different cultures. As such, this chapter provides yet another perspective on sustainable luxury, within a country context that currently lacks research.

Staying within the context of Saudi Arabia, Chap. 6 provides insights '*Towards Circular Luxury Entrepreneurship: A Saudi Female Entrepreneur Perspective*, authored by **Rana Alblowi**, (Prince Sattam Bin Abdulaziz University, Saudi Arabia; The University of Manchester, UK) and **Claudia E. Henninger, Rachel Parker-Strak**, and **Marta Blazquez** (The University of Manchester, UK). This qualitative enquiry illustrates the complexities involved in moving towards a more sustainable luxury fashion industry. Aside from terminological issues that have been identified in the previous chapters, the authors highlight that education can be a key barrier to fully capitalise on new opportunities. Moreover, culture is seen as both a barrier and an enabler, with especially younger generations wanting to support sustainable processes, the new generation of fashion luxury entrepreneurs focuses on innovative ways to make this a reality.

In moving continents to an African setting, Chap. 7 is entitled *'Sustainable Supply Chain Process of the Luxury Kente Textile: Introducing Heritage into the Sustainability Framework'* written by **Sharon Nunoo** (The University of Manchester, UK). This chapter focuses on expanding what we mean by sustainability and how this is reflected within the luxury phenomenon. In line with cultural traditions that have been observed in the previous chapters, the author focuses on the Kente cloth—a type of silk and cotton fabric made of interwoven cloth strips in Ghana, Africa. The author outlines that considering the past is vital in fully understanding what sustainability means. Through in-depth content analysis of secondary resources ranging from website articles and news articles, this chapter documents the textile production process and considers its sustainability with heritage and tradition in mind.

In staying with the theme of jewellery and supply chain, Chap. 8 explores *'Canadian Ethical Diamonds and Identity Obsession: How Consumers of Ethical Jewellery in Italy Understand Traceability,'* written by **Linda Armano** and **Annamma Joy** (Faculty of Management, The University of British Columbia, Canada). Through an ethnographic survey of Italian ethical jewellers and representative customers, this study focuses on how the topic of traceability in the Canadian diamond industry is communicated and negotiated within two representative Italian ethical jewellery stores in Milan and Bologna. A key focus of the chapter is on traceability in an often heavily debated area, namely diamonds. The authors argue that narratives of virtue through consumption can lead consumers to feel they have the ability to buy what they believe in, that opening their wallet is in and of itself an act of power.

Chapter 9, *'Sustainability Claims in the Luxury Beauty Industry: An Exploratory Study of Consumers' Perceptions and Behaviour,'* by **Panayiota J. Alevizou** (The University of Sheffield, UK) broadens our understanding of the 'luxury fashion industry' and focuses its attention on the beauty industry, which is often overlooked. Following on from Chap. 8, the author picks up on the theme of traceability by honing in on communication strategies that are often used by luxury retailers/brands, namely eco-labelling. The author provides interesting consumer insights into a topic that has received increased attention, not least due to having been highlighted as a solution to inform conscious consumers about products that are more 'sustainable' than others in the same product category. As a result five behavioural patterns were identified that encapsulate consumers' relationship with their luxury beauty and its sustainability signals.

Chapter 10, *'What Do You Think? Investigating How Consumers Perceive Luxury Fashion Brand's Eco-labelling Strategy,'* by **Shuchan Luo, Aurelie Le Normand, Claudia E. Henninger**, and **Marta Blazquez** (The University of Manchester, UK) stays within the remit of eco-labelling by exploring Gen Y consumers' perceptions of them within the luxury fashion industry. The findings are surprising in that there is an interesting dynamic emerging between when consumers accept and reject eco-labels and how they unconsciously impact them. The authors outline that whether consumers are aware of eco-labels used almost plays a secondary role; what is more impactful is whether consumers trust the brand. Thus, if a luxury brand is trusted, consumers automatically perceive the eco-label as positive and are convinced this brand is doing their bid for sustainability.

Chapter 11, *' "Take a Stand" the Importance of Social Sustainability and Its Effect on Generation Z Consumption of Luxury Fashion Brands,'* by **Helen McCormick** (Manchester Metropolitan University, UK) and **Pratibha Ram** (The University of Manchester, UK) takes on a more daring tone and moves away from an environmental to a social sustainability focus. The chapter moves us towards the digital age and analyses the importance of 'Taking a Stand' to communicate social sustainability regarding societal issues and its effect on Generation Z consumption of luxury fashion brands.

Chapter 12, *'Exploring Chinese Consumer Attitudes Towards Second-Hand Luxury Fashion and How Social Media eWoM Affects the Decision-Making Process,'* by **Rosy Boardman, Yuping Zhou**, and **Yunshi Guo** (The University of Manchester, UK) provides a nice link between Chaps. 8–11 in that it focuses on issues concerning traceability, whilst at the same time provides insights into social media, and thus is connected with our increasingly digital age. The authors focus on the Chinese consumer context and thus provide a further perspective on second-hand luxury, which has not previously emerged in this edited book. The chapter carefully interlinks the impact eWoM has on the decision-making process when purchasing second-hand luxury fashion. The authors provide some interesting conclusions in regard to wariness of the credibility of eWoM. It was highlighted that the likelihood of impulse purchasing diminished the more eWoM there was. This suggests there may be regional and cultural differences regarding attitudes towards second-hand luxury fashion with more fears about counterfeiting amongst Chinese consumers.

Last, but certainly not least, Chap. 13, '*The Rise of Virtual Representation of Fashion in Marketing Practices: How It Can Encourage Sustainable Luxury Fashion Consumption*,' by **Shuang Zhou, Eunsoo Baek**, and **Juyeun Jang** (The Hong Kong Polytechnic University, HK) provides a conceptual framework addressing key issues emerging with increased digitalisation. The authors highlight that the luxury fashion industry has undertaken initiatives to increase sustainability by employing digital technologies to virtually represent fashion in marketing activities. Within the chapter the authors propose that virtual representation of fashion (VRF) could be used to encourage sustainable luxury fashion consumption (SLFC). This chapter concludes with providing key areas of research and practical implication.

1.3 Final Words

What probably becomes apparent from the content of the edited volume is that that sustainable luxury is a fascinating context that can be explored from a variety of different angles. Seeing as digitalisation brings forward changes in the environment, it will be interesting to see how the future of sustainable luxury will shape.

Finally, we want to say **Thank you!** to all our authors and contributors to this edited book. As outlined in the preface, this edited volume is rather special, as it was in the making for a very long time and has finally materialised. Yet, this would not have happened if it was not for all our wonderful authors, who have dedicated a lot of time!

References

Athwal, N., Wells, V., Carrigan, M., & Henninger, C. E. (2019). Sustainable luxury marketing: A synthesis and research agenda. *International Journal of Management Review, 21*(4), 405–426.

Bala, D. (2021). Do you already own luxury fashion's next it piece? *Vogue* (online). Retrieved February 19, 2022, from https://www.vogue.co.uk/fashion/article/luxury-upcycling

Beauloye, F. E. (2021). Luxury resale: A secondhand strategy for brands. *Luxe Digital* (online). Retrieved February 19, 2022, from https://luxe.digital/business/digital-luxury-reports/luxury-resale-transformation/

Blazquez, M., Henninger, C. E., Alexander, B., & Franquesa, C. (2020). Consumers' knowledge and intentions towards sustainability. *Fashion Practice, 12*(1), 34–54.

Brydges, T., & Hanlon, M. (2020). Garment worker rights and the fashion industry's response to COVID-19. *Dialogues in Human Geography, 10*(2), 195–198.

Brydges, T., Retamal, M., & Hanlon, M. (2020). Will COVID-19 support the transition to a more sustainable fashion industry? *Sustainability: Science, Practice and Policy, 16*(1), 298–308.

Caïs, C. (2021). Is sustainability the next frontier for luxury brands. *Forbes* (online). Retrieved February 19, 2022, from https://www.forbes.com/sites/forbesagencycouncil/2021/11/24/is-sustainability-the-next-frontier-for-luxury-brands/?sh=7036867b96b5

Chevalier, M., & Mazzalovo, G. (2012). *Luxury brand management: A world of privilege* (2nd ed.). Wiley.

Davies, I., Oates, C. J., Tynan, C., Carrigan, M., Casey, K., Heath, T., Henninger, C. E., Lichrou, M., McDonagh, P., McDonald, S., McKechnie, S., McLeay, F., O'Malley, L., & Wells, V. (2020). Seeking sustainable futures in marketing and consumer research. *European Journal of Marketing, 54*(11), 2911–2939.

Davis, J. (2020). How are luxury retailers taking on sustainability? *Harper's Bazaar* (online). Retrieved February 19, 2022, from https://www.harpersbazaar.com/uk/fashion/fashion-news/a29386990/luxury-retailers-sustainability/

Dickson, M., & Warren, H. (2020). A look at labor issues in the manufacturing of fashion through the perspective of human trafficking and modern-day slavery. In S. B. Marcketti & E. E. Karpova (Eds.), *The danger of fashion*. Bloomsbury.

Elkington, J. (2004). Enter the triple bottom line. In A. Henriques & J. Richardson (Eds.), *The triple bottom line, does it all add up?* (pp. 1–16). Earthscan.

Fish, I. (2020). How Covid turned the spotlight on sustainability. *Drapers* (online). Retrieved February 19, 2022, from https://www.drapersonline.com/insight/analysis/how-sustainability-became-a-covid-survival-strategy

Fletcher, K., & Tham, M. (2019). Earth logic- fashion action research plan. *Earth Logic* (online). Retrieved July 18, 2021, from https://earthlogic.info/wp-content/uploads/2021/03/Earth-Logic-E-version.pdf

Granskog, A., Lee, L., Magnus, K. H., & Sawers, C. (2020). Survey: consumer sentiment on sustainability in fashion. *McKinsey* (online). Retrieved February 19, 2022, from https://www.mckinsey.com/industries/retail/our-insights/survey-consumer-sentiment-on-sustainability-in-fashion

Henninger, C. E., Alevizou, P. J., & Oates, C. J. (2016). What is sustainable fashion? *Journal of Fashion Marketing & Management, 20*(4), 400–416.

Henninger, C. E., Alevizou, P. J., Goworek, H., & Ryding, D. (Eds.). (2017). *Sustainability in fashion: A cradle to upcycle approach*. Springer.

Henninger, C. E., Brydges, T., Jones, C., & Le Normand, A. (2022). That's so trashy – Artification in the luxury fashion industry. In A. Joy (Ed.), *Future of luxury: Sustainability and Artification*. de Gruyter.

Iran, S., Joyner Martinez, C. M., Vladimirova, K., Wallaschkowski, S., Diddi, S., Henninger, C. E., McCormick, H., Matus, K. Ninimäki, K., Sauerwein, M., Singh, R., & Tiedke, L., (2022) When mortality knocks: Pandemic-inspired attitude shifts towards sustainable clothing consumption in six countries. *International Journal of Sustainable Fashion and Textiles* (forthcoming).

Kapferer, J. N., & Bastien, V. (2009). *The luxury strategy: Break the rules of marketing to build luxury brands*. Kogan Page.

Keller, K. L. (2009). Managing the growth trade-off: Challenges and opportunities in luxury branding. *Journal of Brand Management, 16*(5/6), 290–301.

Li, G., Li, G., & Kambele, Z. (2012). Luxury fashion brand consumers in China: Perceived value, fashion lifestyle, and willingness to pay. *Journal of Business Research, 65*(10), 1516–1522.

Luo, S., Henninger, C. E., Le Normand, A., & Blazquez, M. (2021). Sustainable what…? The role of corporate websites in communicating material innovations in the luxury fashion industry. *Journal of Design, Business and Society, 7*(1), 83–103.

Marriott, H. (2020). Could the Covid pandemic make fashion more sustainable? *Guardian* (online). Retrieved February 19, 2022, from https://www.theguardian.com/world/2020/dec/28/could-the-covid-pandemic-make-fashion-more-sustainable

Niinimäki, K. (2018). *Sustainable fashion in a circular economy.* Aalto ARTS Books.

Okonkwo, U. (2007). *Luxury fashion branding: Trends, tactics, techniques.* Palgrave Macmillan.

Phau, I., & Prendergast, G. (2000). Consuming luxury brands: The relevance of the 'rarity principle'. *Journal of Brand Management, 8*(2), 122–138.

Ricchetti, M., & De Palma, R. (2020). Will COVID-19 accelerate the transition to a sustainable fashion industry? *UNIDO* (online). Retrieved February 19, 2022, from https://www.unido.org/stories/will-covid-19-accelerate-transition-sustainable-fashion-industry

Rooks, T. (2021). Demand for secondhand luxury goods on the rise. *DW* (online). Retrieved February 19, 2022, from https://www.dw.com/en/online-sales-of-real-authentic-luxury-second-hand-goods-are-growing/a-60029651

Ryding, D., Henninger, C. E., & Blazquez Cano, M. (Eds.), (2018). *Vintage luxury fashion: Exploring the rise of secondhand clothing trade.* Palgrave Advances in Luxury Series. London: Palgrave.

Statista. (2022). Luxury fashion. *Statista* (online). Retrieved February 19, 2022, from https://www.statista.com/outlook/cmo/luxury-goods/luxury-fashion/worldwide

Turunen, L. L. M. (2018). Concept of luxury through the lens of history. In *Interpretations of luxury.* Palgrave advances in luxury. Palgrave Macmillan.

Turunen, L. L. M., & Leipämaa-Leskinen, H. (2015). Pre-loved luxury: Identifying the meanings of second-hand luxury possessions. *Journal of Product & Brand Management, 24*(1), 57–65.

Turunen, L. L. M., Cervellon, M. C., & Carey, L. D. (2020). Selling second-hand luxury: Empowerment and enactment of social roles. *Journal of Business Research, 116*, 474–481.

Wells, V., Athwal, N., Nervino, E., & Carrigan, M. (2021). How legitimate are the environmental sustainability claims of luxury conglomerates? *Journal of Fashion Marketing and Management, 25*(4), 697–722.

2

The Hidden Value of Second-Hand Luxury: Exploring the Levels of Second-Hand Integration as Part of a Luxury Brand's Strategy

Linda Lisa Maria Turunen and Claudia E. Henninger

2.1 Introduction

The booming resale market has put a spotlight on the second-hand phenomenon, which has also seen an increase within the luxury market. According to the Boston Consulting Group (2019), luxury resale is expected to grow four times faster than the primary luxury market—12% per year as compared to 3% growth. It may not be surprising that various luxury brands have capitalized on this growing market segment. For example, the Kering Group, a global luxury conglomerate, invested in Vestiaire Collective, a platform for pre-owned luxury goods, and acquired a stake of approximately 5% from the resale titan. At the same time the

L. L. M. Turunen (✉)
Aalto University, Espoo, Finland
e-mail: linda.turunen@aalto.fi

C. E. Henninger
Department of Materials, The University of Manchester, Manchester, UK

Kering Group has also taken up a seat as a board member in the Vestiaire Collective (Kering, 2021). The Kering Group is only one of many luxury brands that have explored ways to capture resale opportunities, finding control of their market presence and to enhance the value for their customers, without sabotaging the primary market.

Luxury and designer-led fashion brands with high-quality products are desired in the growing second-hand market (ThredUp, 2021). There are multiple reasons as to why the resale phenomenon has taken up: first and foremost, consumers, who are actively taking steps towards extending the lifecycle of their used products through selling them, whilst at the same time making money of this transaction (Turunen et al., 2020). Second, we see companies emerging, who tap into this market, by providing platforms that facilitate these resale opportunities and thus bring value for consumer-to-consumer transactions, either by enhancing the ease to find specific luxury garments, by making the resale process more convenience or by enhancing trust in transactions between strangers by offering authentication services (Elle, 2021). Last, the dominant zeitgeist of sustainability and the circular economy have fostered a rise in resale (Sun et al., 2021). Resale extends the product's lifetime or multiplies the usage time of the product, and thus can counter act the wastefulness of resources the industry has been accused of. Sustainability discussions have also positively boosted the social acceptance around second-hand consumption (Amatulli et al., 2018). During the past years, the stigma associated with purchasing pre-owned goods has diminished, and it has become even trendy (Gorra, 2017). The shifting attitudes of consumers and renewed acceptability have greatly influenced the growth of not only the second-hand but also the resale market (e.g., Ferraro et al., 2016; Beard, 2008).

One would expect that luxury and designer-led fashion brands would be eager to engage in the second-hand market to control their brand's image, product availability and service experience, make profits and even enhance sustainability perception. Yet, despite second-hand activities growing at a steady pace, many brands seem to have left the market of their desired treasures for resale marketplaces. This chapter examines second-hand markets from brands' perspectives and sheds light on how the luxury resale market has and still is evolving. The chapter aims to clarify and map second-hand strategies brands have applied as well as build an understanding surrounding the hidden value resale actions may contain. The topic will be examined especially in the light of sustainability.

The chapter will proceed as follows: first, the previous literature about second-hand and resale activities as a service will be reviewed. Second, the alternative paths of integrating second-hand activities under the brand will be explored by analysing existing business models in the market. Finally, the chapter will conclude the alternative ways of incorporating and utilizing second-hand market as a part of brand building of luxury- and designer-led premium brands.

2.2 Literature Review

Second-hand refers to things that are not new and have been owned by someone else (Oxford Dictionary, 2021). The existence of the second-hand market and the subsequent products being on offer are dependent on primary market offerings and first-hand consumer's willingness to dispose of these products (Turunen et al., 2020). Thus, customers may take active and emerging roles when becoming suppliers as well as co-producers of value (Hvass, 2015; Turunen et al., 2018). To enhance and ease secondary transactions, resale platforms have emerged, which take on the role of the facilitator between sellers and buyers, as well as deal with aspects surrounding a reverse supply chain. Within the last couple of years, the resale arena is getting busier with startups emerging, who help already-existing brands to integrate the second-hand services within their company portfolio (BoF, 2021a). This chapter focuses specifically on luxury and designer-led fashion brands' endeavours in prolonging their products' life through resale activities.

2.2.1 Second-Hand and Global Resale Platforms/ Digitalization

Although second-hand retailing dates back many centuries (Hvass, 2015), the ways items are bought and sold are continuously evolving. The luxury and fashion resale market is busy with global and local players: the fast development of online channels—which offer space for peer-to-peer marketplaces, hosted apps dedicated to second-hand apparel and accessories and specialized online platforms—confers worldwide visibility to pre-owned luxury goods. The Boston Consulting Group (2019) even suggested that four out of five people participate in the second-hand luxury market, by not only getting informed about offerings but also actively trading online. The biggest resale players in the arena of luxury fashion are: the US based The RealReal with 24 million users in across the world (The RealReal, 2021), and the European-based Vestiaire Collective pushing the total to almost 10 million members worldwide (Vestiaire Collective, 2022).

Resale platforms, such as The RealReal and Vestiaire Collective, are businesses focusing on orchestrating the resell process of previously owned luxury and designer-led fashion goods. The resale market has its own logic and dynamics, and high-quality luxury goods are just items traded in the platform. From a luxury brand's perspective, resale opportunities may also be seen as problematic, as they are uncontrollable (BoF, 2021b). For example, luxury brands cannot influence what products are available (e.g., decade old, licensed pieces or rare limited editions) and their quality/condition (e.g., nearly brand-new or worn and patinated), whether the products are authentic, how the products are represented on the platform or what kind of service experience the resale platform offers.

Resale platforms are serving two kinds of luxury customers, those who sell luxury goods and those who buy reused luxury artefacts (Turunen et al., 2020). While platforms are scaling up the accessibility of pre-used luxury goods, it has not been always straight forward: one example is the long legal battle between Chanel and The RealReal, which concerns a trademark infringement and counterfeiting. Chanel is accusing The RealReal selling counterfeit Chanel handbags, which of course can be problematic (The Fashion Law, 2018, 2021).

The extensive growth of the resale market and uncontrollability has also encouraged brands to examine ways of being involved in the resell process of their own goods. Bridging primary and secondary markets, strategic alliances and partnerships between companies have emerged (BoF, 2021c). Besides creating mutually beneficial collaborations with resale platforms, a brand that is operating in the primary market can facilitate business model adaptions towards sustainability, for example, by offering second-hand as a service for their own brand's products.

2.2.2 Second-Hand as a Service

While second-hand as a concept is nothing new, and even global luxury resale platforms have been building their businesses for nearly a decade, luxury- and designer-led fashion brands' engagement with the end-of-life of their products is a more recent phenomenon (BoF, 2021c).

A brand who operates in the primary market may adopt second-hand or resale services—a specific form of product-service system—as part of the brand's operations, if pursuing to extend the lifetime of their products. Tukker (2004: 246) defines a product-service system as consisting of "tangible products and intangible services designed and combined so that they jointly are capable of fulfilling specific customer needs." Product-service systems can create new sources of added value for consumers and boost brand competitive-position in the market, since they can enhance customer loyalty and help to build a unique relationship with their clients (Tukker, 2004). Although product-service systems may have the potential to extend a product's lifetime and minimize resources used throughout the product's lifecycle, this is not necessarily always the case (Kjaer et al., 2019). It is argued that the convenience and ease of resell may even encourage to speed up the consumption cycle: the Boston Consulting Group found that 62% of the products sold on second-hand luxury platforms are either unworn or scarcely worn (BCG, 2019).

Contextualizing Tukker's (2004) idea of product-service systems in the second-hand market, Hvass (2015) pointed out two concrete strategies to implement downstream value chains in the textile industry: (1) fashion brands could facilitate take-back schemes for fibre recycling purposes, with the actual recycling being organized by a third party. Here, the value is in the volume of the material, thus it may not be surprising to see that these take-back schemes are linked to monetary incentives, which encourages consumer to purchase more and new items. (2) Fashion brands could offer take-back services accompanied with resell services for prolonging the life of their own products. In this case, the value in the garment or product itself is acknowledged, and it might be resold or upcycled. Various tactics source the pre-used products, such as incentives, are often used. In terms of luxury and premium brands, the item's post-retail value is often so high that recycling of the actual product is rarely the case. However, it is valuable to examine what these strategies require when implemented as part of the brand's offering.

Second-hand requires a so-called reverse service supply chain (Beh et al., 2016), which means resale has its own logic and thus is different to the primary market. Reverse services are often complex, because the output consists of a bundle comprising tangible and intangible components (Davis & Heineke, 2003). The primary market follows a first-order supply chain which requires sourcing raw materials and manufacturing the product, but reverse logistics include, for example, collecting garments (e.g., take-back), sorting, re-distribution and pricing of pre-used products (Hvass, 2015). Thus, second-hand as a service may sound attractive for consumers, but for brands operating in the primary market, it would require completely new and separate operating model.

Previous literature suggests several benefits that the engagement with post-retail activities may provide for the brand. For example, it gives an opportunity to strengthen the relationship with existing customers and even to reach new market segments (Hvass, 2014, 2015; Tukker, 2004). Yet, as previously alluded to, it can also bear challenges and tarnish a brand's brand image (BCG, 2019), as it cannot always be guaranteed what is available in the market, nor the condition. If selling through third-party platforms, there are also issues surrounding authenticity of products and the subsequent sale of potential counterfeit products, which can be harmful for luxury and designer-led fashion brands (e.g., The Fashion Law, 2018, 2021).

Further, second-hand has also become a way of embracing sustainability (Carrigan et al., 2013), by extending a product's lifetime with multiple owners. Klepp et al. (2020) analysed clothing lifespans and argued that if second-hand products are replacing the acquisition of a new item, it may be regarded to have a smaller environmental impact. While second-hand is often seen as a sustainable choice, it needs to be acknowledged that the change of ownership is rarely environmentally neutral when compared with products used multiple times by the same owner:

there might be activities related to reverse supply chain, such as transportation, additional laundering, sorting and storing involved, all of which can have an environmental impact (Fisher et al., 2011; Farrant et al. 2010; Botticello, 2012). However, if the change of ownership enables increasing numbers of usage times for artefacts, the relative environmental impact of the item will be smaller the longer the lifespan is (Sandin & Peters, 2018; Fisher et al., 2011).

2.3 Materials and Methods

This chapter examines the second-hand market and aims to capture the dominant zeitgeist of the resale market, specifically from luxury and designer-led fashion brands' perspective. The empirical data consists of two kinds of secondary data collected: luxury and fashion brand's communication (websites) and media presence and press releases. The guiding questions for data exploration were: (1) what is going on in the resale market, which luxury and fashion brands are involved and how, (2) what kinds of discussions relate to second-hand involvement—how is the resale market presence validated, argued for and against.

Empirical data collection started with market screening to build an understanding of the existing ways on how luxury and designer-led fashion brands are involved in the resale market. The mapping was conducted by screening through luxury companies' (n = 35) and luxury resale platforms' (n = 30) webpages. The number of resale platforms is ever increasing, and despite the global presence of the largest platforms, most of the resale-platform companies are operating locally. The underlying assumption was that if the brand is involved in second-hand activities or offering a second-hand service, it is publicly announced on the webpage. The findings of the screening showed that luxury brands were rarely executing these activities themselves, but rather collaborating with resale platforms. To expand the scope to cover different ways of second-hand integration, the search was extended to cover premium fashion brands, multi brand-retailers with luxury and high-end brands and department stores.

The first step was further supported by conducting keyword search with specific brand names (emerging from the first step of empirical data collection) and combining keywords such as second-hand, resale, pre-used, pre-owned, vintage, luxury and fashion. Especially fashion-related magazines and international newspapers written in English were screened to add aid as additional support.

Uncovering the strategies luxury and designer-led fashion brands applied to be involved in second-hand market, the collected secondary data was analysed by the means of content analysis (Belk et al., 2013). First, it was observed what the existing actions brands are doing and how they are integrating second-hand in the business models. After mapping the existing actions, they were categorized into overarching themes or strategies. The magazine articles were analysed in greater detail within the categories to uncover the reasonings for the chosen strategies and aid as additional support.

2.4 Findings

Luxury items are available in marketplaces that brands cannot control, and new, emerging target audiences as well as current luxury customers are interacting with them. As already alluded to, there are various ways brands can engage with the pre-owned market to capture the value, stay relevant for and connect with their customers. Next the alternative approaches how luxury and designer-led fashion brands are implementing second-hand or resale activities in their operations will be mapped.

Various resale- and second-hand-related activities that have been applied by luxury and designer-led fashion brands were categorized and unveiled three main strategies of second-hand integration: luxury brands can be

1. collaborating with a resale player;
2. integrating the second-hand services with the help of third-party actors; or
3. building a full-fledged second-hand service in-house.

The level of integration, responsibility and control a brand differs between the different strategies. Table 2.1 summarizes the characteristics of brand's second-hand strategies.

What becomes apparent from Table 2.1 is that there seems to be a spectrum from low to high brand control and economic risk. High brand control and integration is resource intensive path with own inventory (high economic risk), but the control enables facilitating the branded second-hand experience. For example, the greatest amount of control and the highest economic risk for retailers is in option 3, the "in-house" option, whilst the lowest economic risk and lowest amount of control can be seen in option 1, collaborations. Collaboration may boost visibility for both parties, but the second-hand experience is solely in the hands of the resale player. This is further explored in the next section.

2.4.1 Second-Hand Integration

The level of integration is differentiating among the strategies. While the *collaboration* arena represents a separate, mutually beneficial agreement with resale platforms, it also stays outside of the brand's core business. A collaborative strategy offers the least control and is also the least resource intensive for the brand. One reason as to why it is less resource intensive is the fact that the reverse supply chain with its unique characteristics is managed by the resale platform, whereas the luxury brand is only involved in a philanthropical manner (see Halme & Laurila, 2009). Various luxury brands, including Stella McCartney, Burberry and Gucci, have partnered with the luxury resale platform The RealReal (Vogue, 2020). The brands have different kinds of partnerships and services co-created with the resale platform. The activities seem to serve especially the customer-seller, and the existing luxury client: the partnership aims to boost the new sales by either offering shopping credits or inviting customer-sellers for exclusive experience with the luxury brand. On the other hand, Vestiaire Collective is focusing on "brand approved" programme to involve selected brands in the take-back and authentication processes.

Premium and designer-led fashion brands had campaign-type examples with small and local resale players. For example, in 2019 Marimekko

Table 2.1 Three strategies of second-hand integration in luxury resale

	Collaboration with resale player	Second-hand as an integrated offering	Full set of second-hand service in-house
Description	Brand collaborates with a resale platform, redirected to third-party website. The resale actions are completely managed by a resale platform.	Brand is involved facilitating the exchange of their pre-used goods. Branded resale process is organized or supported by third party.	Second-hand/take-back is an added service or programme offered by the brand. Brand buys/incentivizes back its own products and sells them.
Responsibility of the process	Brand is not involved in resale activities No inventory	Third party facilitates the process under brand No inventory	Brand owns the product and facilitates the process Inventory
Level of integration	Outside of the core business	Integration with the core business	Extension of the core business
Brand's control	Low	Medium	High
Examples	Global brands who are not building the resale programme under the brand, for example, Stella McCartney & The RealReal, Alexander McQueen & Vestiaire Collective.	Big-scale brands who want to own their second cycle. For example, brands like Patagonia, Levi's, Eileen Fisher and Lululemon work with Trove. Additional third-party models offered, for example, Reflaunt and The Archivist. Third party offers the white cloves branded service with technology.	Smaller brands who have resources to build the second-hand programme inside the brand, for example Marimekko, Mulberry, Samuji and Arela.

partnered with the local resale platform Vestis and opened collection points in Marimekko's own brick-and-mortar stores. The resale-partner maintained, quality checked, priced and sold the previously used Marimekko items. In addition, brands such as Reformation, Madewell and Rent the Runway are partnering with ThredUp who organizes the resell or recycling of used goods. Collaborations with resale players offer visibility for both partners. New forms of strategic collaborations might be expected among luxury brands, since the global luxury group, Kering acquired 5% stake of Vestiaire Collective in spring 2021 (Vogue, 2021a). The investment brought Kering to Vestiaire Collective's board and gave a front seat to observe where the resale market is heading.

If collaborations form on one end of the spectrum, that can be described as "light touch," *full-fledged in-house second-hand service* provides the opposite end to this spectrum. In other words, a brand offers take-back and resell services for its own items. It represents the extension of a brand's core business, and offers the most control, yet requires the most resources to operate. While the in-house service requires adaption of a reverse service supply chain—everything from sourcing, maintaining, repairing and selling—it turned out to be a strategy especially for small-scale and innovative brands who are devoted to altering their business models towards more sustainable practices. Various small-scale designer-led fashion brands such as Samuji and Arela are offering take-back and resell services for their own goods. The brands produce high-quality pieces with timeless designs and are committed to lengthen the lifecycle of their own pieces. Also, Mulberry offers selection of maintained and repaired pre-loved handbags, and Marimekko launched a *Pre-loved* collection in-house in 2021, which included a curated collection of 60 vintage pieces from the 1960s to the 2000s.

A relatively recent phenomenon, which can be situated in the middle of the continuum and thus between collaborations and in-house second-hand services, is the *second-hand as an integrated offering*. In this branded second-hand strategy, the resale process is positioned with the brand portfolio; however, the service is organized and supported by a third-party player, who manages the reverse supply chain. New emerging companies are positioning themselves as branded second-hand service providers who offer a white-label service for brands. The companies have different approaches and emphasis the service aspect. For example, Trove, a well-established resale company, has built resale platforms for brands like Eileen Fisher, Levi's and Patagonia. Trove takes care of the whole buying and selling process, covering the distribution centres and handing logistics.

The Archivist offers data-driven resale-as-a-service technology, and differently to Trove, they target luxury brands (The Archivist, 2021). The Archivist offers various models from peer-to-peer marketplace facilitation to more controlled consignment models, which manage the inventory (BoF, 2021a), and by doing so they suggest that it brings value to brands by monetizing on their brand history and leveraging the vintage trend (LVMH, 2020). Further, data-driven The Archivist offers data dashboards to provide insights for brands, and to build their understanding of the industry dynamics and developments of their own products in resale market (The Archivist, 2021; VogueBusiness, 2021). Despite The Archivist claiming to be targeting luxury brands, so far concrete results (e.g., luxury brand partnerships) are not found in the empirical data. For now, it seems that most of the brands following the "second-hand as an integrated offering" strategy are large-scale, premium fashion brands.

A further example is Reflaunt, which offers a technology that connects brands to the global network of second-hand marketplaces. It aims to ease and speed the re-selling process. The company is backed by Kering and works with companies such as Balenciaga, Ganni and Jimmy Choo (VogueBusiness, 2021; Reflaunt, 2021).

2.4.2 Retailers' Resale Services

The empirical exploration of luxury and designer-led fashion brands resale also unveiled the emerging number of retailers and multi-brand online stores who have launched their own resale marketplaces. For instance, Farfetch offers a *Second Life* service for pre-loved luxury handbags (Farfetch, 2019), and recently also partnered with ThredUp who offers resale-as-a-service technology (VogueBusiness, 2021). Mytheresa partners with Vestiaire Collective by offering a designer resale programme to Mytheresa high-end luxury customers (Businesswire, 2021). In addition, Zalando has extended their offering by launching Zalando Zircle, through which consumer-sellers can gain credits to shop in Zalando (Zalando, 2021).

The retailer's resale-services aim to turn the customer-sellers into new primary consumers of the retail-platform, by offering short-term monetary benefits for future purchases. Multi-brand retailers who have included resale opportunities in their store offer credit to purchase new items. Thus, the expectation is that consumers, who dispose and sell, become first-hand customers in the primary market. Retailers' resale-services pursue to keep customers in the loop and possibly boost customer loyalty. This logic is aligned with Boston Consulting Group's (2019) finding which suggest that around 30% of respondents' primary reason for selling was to purchase new goods.

It is noteworthy to point out that department stores are also following the resale trend: the UK-based Selfridges opened its doors for Vestiaire Collective, who launched its first in-store destination with Selfridges (Selfridges, 2019), containing Vestiaire Collective's expertly curated and authenticated pre-loved luxury pieces. Further, in partnership with Vestiaire and Depop, Selfridges launched resale service called RESELLFRIDGES (Retail Gazette, 2021; RESELLFRIDGES, 2021). Neiman Marcus, a high-end department store in the United States, is collaborating with Fashionphile, of which it has also acquired a minority stake in 2019 (Bizwomen, 2021). The British department store Harvey Nichols partnered with resale technology provider, Reflaunt to launce resale service for their customers (BoF, 2021d).

2.4.3 Collaboration with Resale Platform Serves the Existing Luxury Clients

Despite luxury items being widely transacted in global resale platforms or monetized in the retailer's resale marketplaces, it became evident that luxury brands are rarely involved in full-fledged second-hand services by themselves. Based on the data, the integration of second-hand services is on a collaborative level. Sustainability pioneer, Stella McCartney was the first luxury brand collaborating with the luxury consignment platform,

The RealReal. The collaboration has been ongoing since 2017, and consigning customer receives $100 credit to shop the brand (The RealReal, 2022). The partnership brings value especially for Stella McCartney's existing customers, those who are willing to dispose of their items. The incentive offered for consigning customers aims to keep the customer returning.

Burberry has been partnering with The RealReal, since 2019. They encourage their customers to consign Burberry goods on their resale platform and offer them (consigning customers) a personal shopping experience at Burberry that includes high tea in 18 US stores (VogueBusiness, 2019). Sourcing items for resale has also turned Alexander McQueen's focus on their existing customers: McQueen has contacted their best clients and offers store credit in return for accepted artefacts from previous collections (Vogue, 2021b; Forbes, 2021). Alexander McQueen is involved in the sourcing and authentication processes of their own pieces, but the resale is organized by Vestiaire Collective.

Based on the empirical exploration, there is currently no luxury brand that is involved in selling their own previously used or owned pieces. The luxury brands that are active in the resale arena and second-hand related services are predominantly following a collaboration approach with resale platforms as described earlier. Thus, luxury brands keep the focus on their core business yet boost customer loyalty of existing customers by offering additional services with branded experience. Further, the collaboration may secure the authenticity of the selected luxury pieces in the resale market. Based on the findings, luxury brands keep their exclusivity by not directly serving second-hand customers.

2.4.4 Branded Second-Hand Experience

Differently to luxury brands who are collaborating with global resale platforms to find mutually beneficial strategic alliances, the empirical data unveiled that designer-led and premium fashion brands are engaged with the end-of-life of their products and were actively building a branded second-hand experience, also for second-hand buyers. The applied strategies are aligned with Hvass (2015) findings concerning strategies implementing downstream value chain in the textile industry.

To build upon Hvass's (2015) product's end-of-life strategies, the empirical data indicated that there are two ways of implementing the branded second-hand experience under the brand's offering: (1) second-hand services operated with the help of third-party player or (2) full-fledged services organized independently in-house. Which one of these strategies were applied, seemed to be dependent on the available of resources and their scalability, and thus, the size of the company: large global brands seemed to scale up the branded resale experience with the help of third-party contractors, while small-sized fashion brands with manageable volumes fully integrated the service as their core business offering. Offering the integrated branded resale experience for both sellers and (second-hand) buyers enables brands to engage existing customers and brand communities, but also expand to new customer segments.

In terms of luxury brands, the branded experience extends only to existing customers or luxury product owners, who may turn to become customer-sellers. Luxury brands are not serving the second-hand buyer market with their own branded products. This limitation might help to control the luxury experience, but also neglects luxury customers who are in the resale arena for older treasures that may be not available in the primary market anymore. The purchase experience of pre-used luxury items—whether they are rare old vintage pieces or masstige products—is outsourced for collaboration partner, such as resale platform. Luxury brands with resale-platform collaborations are offering services—such as ease the process of disposal by collecting the items, offer incentives to motivate to dispose and possibly purchase new, or provide in-house experience—to all customers who possess their item, whether it has been acquired first- or second-hand originally. In sum, luxury brands are limiting their services only for customers who possess their artefact, regardless of how the customer became the owner of the artefact.

2.4.5 Hidden Value of Second-Hand Luxury

The empirical data exemplified the close linkage of second-hand and sustainability. Sustainability pursuits the long lifecycle of the products were increasingly highlighted in media and by brands involving in the resale. However, the product's lengthened lifespan with multiple users might be a step towards more sustainable fashion future, yet it is not a solution for industry's sustainability problems. If the resale market is used only to boost sales of brand-new items—as some incentivized offerings and collaborations may support—the fashion industry's shift towards sustainable practices is not happening until the primary market changes (BoF, 2020). As its best, integrating second-hand activities in the business model might give a possibility to expand and reform the brand's offering and even find ways to replace some of the brand-new items in the future.

Despite the sustainability aspects the resale market offers valuable information and pathways forward for luxury brands. Partnering with resale platforms may boost the sustainability aura, increase brand loyalty of existing customers, encourage customer-sellers to buy first-hand (if incentivised) or enable the fight for authenticity.

Bringing second-hand services in-house, enables brands to regain power over how they are presented in the market and possibly even create new revenue streams as well as expand their target audience from primary to secondary customers. Further, integrating the second-hand activities in-house may offer valuable information for product development and control over resale pricing and experience. Figure 2.1 summarizes the key findings of different resale strategies.

2.5 Conclusions

Luxury brands are building a mutually beneficial partnership with global resale platforms to keep the eye on the market and to serve their existing clients. While focusing on the core business and luxury experience, the luxury brands are not closely involved in the resale process of their own artefacts, instead, resale platforms have the responsibility of the sales

2 The Hidden Value of Second-Hand Luxury: Exploring...

Fig. 2.1 Three main strategies of second-hand integration for luxury brands

process and second-hand buyers. Simultaneously new emerging startups have entered to ease the process of building successful second-hand programmes especially for premium brands who have the scale and brand community to tap into. Some forward-thinking small brands are even figuring out their own way of constructing the resale offering.

Second-hand as a service may support the sustainability pursuits by circulating items and extending the product lifecycles. To overcome the fear of cannibalizing sales of new products, the incentivizing actions that boost the sales of new items have often adopted. Despite the circulating products and engaged customers are the connecting links bridging the primary and secondary market, the logics of the resale market has its own characteristics. Thus, brands should value and regard the resale market as an opportunity of its own. Only time will show how luxury brands' deepening relationships with global luxury resale platforms will develop or how emerging innovative data-driven startups will shift the role of resale in luxury arena.

References

Amatulli, C., Pino, G., De Angelis, M., & Cascio, R. (2018). Understanding purchase determinants of luxury vintage products. *Psychology & Marketing, 35*(8), 616–624.

Beard, N. D. (2008). The branding of ethical fashion and the consumer: A luxury niche or mass-market reality? *Fashion Theory, 12*(4), 447–467.

Beh, L. S., Ghobadian, A., He, Q., Gallear, D., & O'Regan, N. (2016). Second-life retailing: A reverse supply chain perspective. *Supply Chain Management: An International Journal.*

Belk, R. W., Fischer, E., & Kozinets, R. V. (2013). *Qualitative Consumer and Marketing Research.* Thousand Oaks, CA: Sage.

Bizwomen. (2021). *Neiman Marcus extends partnership with luxury reseller Fashionphile.* Published 27.4.2021, Anne Stych. https://www.bizjournals.com/bizwomen/news/latest-news/2021/04/neiman-marcus-extends-partnership-fashionphile.html?page=all

BoF. (2020). *Is resale actually good for the planet?* Published 7.1.2020, Sarah Kent. https://www.businessoffashion.com/articles/sustainability/is-resale-actually-good-for-the-planet

BoF. (2021a). *New resale start-up sees data as the answer to courting luxury brands.* Published 12.5.2021, Cathaleen Chen. https://www.businessoffashion.com/articles/technology/new-resale-start-up-sees-data-as-the-answer-to-courting-luxury-brands

BoF. (2021b). *Business of fashion: For brands, is resale actually worth it?* Published 8.2.2021, Cathaleen Chen. https://www.businessoffashion.com/articles/retail/for-brands-is-resale-actually-worth-it

BoF. (2021c). *The future of fashion resale report—BoF insights.* https://www.businessoffashion.com/reports/retail/the-future-of-fashion-resale-report-bof-insights

BoF. (2021d). *Harvey Nichols partners with Reflaunt to launch resale service.* Published 5.8.2021, Darcey Sergison. https://www.businessoffashion.com/news/retail/harvey-nichols-partners-with-reflaunt-to-launch-resale-service

Boston Consulting Group. (2019). *Why luxury brands should celebrate the pre-owned boom.* https://www.bcg.com/fr-fr/publications/2019/luxury-brands-should-celebrate-preowned-boom

Botticello, J. (2012). Between classification, objectification, and perception: Processing secondhand clothing for recycling and reuse. *Text, 10*(2), 164–183.

Businesswire. (2021). *Mytheresa partners with Vestiaire collective to introduce a unique resale service dedicated to Mytheresa's high-end luxury customers to reinforce the shift to circularity as part of the fashion ecosystem.* Published 9.6.2021. https://www.businesswire.com/news/home/20210608005814/en/Mytheresa-Partners-With-Vestiaire-Collective-to-Introduce-a-Unique-Resale-Service-Dedicated-to-Mytheresa's-High-end-Luxury-Customers-to-Reinforce-the-Shift-to-Circularity-as-Part-of-the-Fashion-Ecosystem

Carrigan, M., Moraes, C., & McEachern, M. (2013). From conspicuous to considered fashion: A harm-chain approach to the responsibilities of luxury-fashion businesses. *Journal of Marketing Management, 29*(11–12), 1277–1307.

Davis, M. M., & Heineke, J. (2003). *Managing services – Using technology to create value.* McGraw-Hill/Irwin.

Elle. (2021). *The rise of the resale market: The landscape inside the fashion industry's new mindset.* Published in 7.5.2021, Fernando Aguileta de la Garza. https://elle.education/business/the-rise-of-the-resale-market-the-landscape-inside-the-fashion-industrys-new-mindset/

Farfetch. (2019). *Farfetch second life: Sell your designer bags.* https://www.farfetch.com/positively-farfetch/secondlife/fi

Farrant, L., Olsen, S. I., & Wangel, A. (2010). Environmental benefits from reusing clothes. *The International Journal of Life Cycle Assessment, 15*(7), 726–736.

Ferraro, C., Sands, S., & Brace-Govan, J. (2016). The role of fashionability in second-hand shopping motivations. *Journal of Retailing and Consumer Services, 32,* 262–268.

Fisher, K., James, K., & Maddox, P. (2011). *Benefits of reuse case study: Clothing* (p. 41). WRAP.

Forbes. (2021). *Partnerships like Vestiaire collective and Alexander McQueen are fueling the circular economy.* Published 18.2.2021, Sharon Edelson. https://www.forbes.com/sites/sharonedelson/2021/02/18/partnerships-like-the-vestiaire-collective-and-alexander-mcqueen%2D%2Dare-fueling-the-circular-economy/?sh=dd8ed9d3ff68

Gorra, C. (2017). *The new normal: Luxury in the secondary market.* https://digital.hbs.edu/innovation-disruption/new-normal-luxury-secondary-market/

Halme, M., & Laurila, J. (2009). Philanthropy, integration or innovation? Exploring the financial and societal outcomes of different types of corporate responsibility. *Journal of Business Ethics, 84*(3), 325–339.

Hvass, K. K. (2014). Post-retail responsibility of garments–a fashion industry perspective. *Journal of Fashion Marketing and Management.*

Hvass, K. K. (2015). Business model innovation through second hand retailing: A fashion industry case. *Journal of Corporate Citizenship, 57*, 11–32.
Kering. (2021). Press release 1.3.2021. https://www.kering.com/en/news/vestiaire-collective-announces-a-new-eur178m-us-216m-financing-round-backed-by-kering-and-tiger-global-management-to-accelerate-its-growth-in-the-second-hand-market-and-drive-change-for-a-more-sustainable-fashion-industry
Kjaer, L. L., Pigosso, D. C., Niero, M., Bech, N. M., & McAloone, T. C. (2019). Product/service-systems for a circular economy: The route to decoupling economic growth from resource consumption? *Journal of Industrial Ecology, 23*(1), 22–35.
Klepp, I. G., Laitala, K., & Wiedemann, S. (2020). Clothing lifespans: What should be measured and how. *Sustainability, 12*(15), 6219.
LVMH. (2020). *LVMH innovation awards*. https://www.lvmh.com/lvmh-innovation-award-2020/the-archivist/
Reflaunt. (2021). Company website. https://www.reflaunt.com
RESELLFRIDGES. (2021). *Resellfridges – sell your preloved accessories with us*. https://resellfridges.com
Retail Gazette. (2021). *Fast fashion retail and resale: A way forward?* Published: 12.4.2021, Sahar Nazir. https://www.retailgazette.co.uk/blog/2021/04/fast-fashion-retail-and-resale-a-way-forward/
Sandin, G., & Peters, G. M. (2018). Environmental impact of textile reuse and recycling–a review. *Journal of Cleaner Production, 184*, 353–365.
Selfridges. (2019). *Selfridges meets Vestiaire collective*. https://www.selfridges.com/GB/en/features/articles/selfridges-meets/vestiaire-collective/
Sun, J. J., Bellezza, S., & Paharia, N. (2021). Buy less, buy luxury: Understanding and overcoming product durability neglect for sustainable consumption. *Journal of Marketing, 85*(3), 28–43.
The Archivist. (2021). Company website. https://thearchivist.com
The Fashion Law. (2018). *Chanel is suing the RealReal for allegedly selling counterfeit bags*. Published 16.11.2018. https://www.thefashionlaw.com/chanel-is-suing-the-realreal-for-allegedly-selling-counterfeit-bags/
The Fashion Law. (2021). *Chanel, the RealReal agree to mediation in escalating counterfeiting, antitrust fight*. Published 9.4.2021. https://www.thefashionlaw.com/chanel-the-realreal-agree-to-mediation-amid-escalating-counterfeiting-antitrust-fight/
The RealReal. (2021). *TRR by the numbers*. Access 18.1.2022 https://www.therealreal.com/about

The RealReal. (2022). *Stella McCartney x the RealReal*. Access 18.1.2022 https://promotion.therealreal.com/stellamccartney/
ThredUp. (2021). *Resale report*. Available at: https://www.thredup.com/resale/
Tukker, A. (2004). Eight types of product–service system: Eight ways to sustainability? Experiences from SusProNet. *Business Strategy and the Environment, 13*(4), 246–260.
Turunen, L. L. M., Leipämaa-Leskinen, H., & Sihvonen, J. (2018). Restructuring secondhand fashion from the consumption perspective. In *Vintage luxury fashion* (pp. 11–27). Cham: Palgrave Macmillan.
Turunen, L. L. M., Cervellon, M. C., & Carey, L. D. (2020). Selling second-hand luxury: Empowerment and enactment of social roles. *Journal of Business Research, 116*, 474–481.
Vestiaire Collective. (2022). *The Smart side of Fashion-report*. https://www.vestiairecollective.com/fashion-report/
Vogue. (2020). *Gucci and the RealReal announce a game-changing partnership*. Published 5.10.2020, Emily Farra. https://www.vogue.com/article/gucci-the-realreal-partnership-secondhand-consignment
Vogue. (2021a). *What does Kering's deal with Vestiaire Collective mean for secondhand fashion—And the entire industry?* Published 9.3.2021, Emily Farra. https://www.vogue.com/article/kering-vestiaire-collective-resale-secondhand-fashion-industry-future
Vogue. (2021b). *Alexander McQueen launches a buy-back program with Vestiaire Collective*. Published 16.2.2021, Steff Yotka. https://www.vogue.com/article/alexander-mcqueen-launches-a-buy-back-program-with-vestiaire-collective
VogueBusiness. (2019). *Burberry's partnership with the RealReal signifies a real shift*. Published: 7.10.2019, Meghan McDowell. https://www.voguebusiness.com/companies/burberrys-partnership-realreal-secondhand
VogueBusiness. (2021). *Startup spotlight: The Archivist puts brands in charge of resale*. Published 6.7.2021, Meghan McDowell. https://www.voguebusiness.com/technology/startup-spotlight-the-archivist-puts-brands-in-charge-of-resale
Zalando. (2021). *Zircle – The wardrobe of the future*. https://corporate.zalando.com/en/magazine/zircle-wardrobe-of-the-the-future

3

Sustainable Eco-luxury in the Scandinavian Context

Kirsi Niinimäki

3.1 Introduction

The Scandinavian lifestyle is well known for its environmental value base and its appreciation of pure nature and slow living (Ollila, 1998). This environmental value base lays the foundations for Scandinavian consumers' understanding of luxury. The hypothesis in this study is that luxury in the Scandinavian context is connected to sustainable values, eco-appreciation, natural-based alternatives, and good 'everyday' design. Scandinavians see eco-luxury as something not meant for wealthy people and elite consumers alone, but as a good, functional design that is based on ecological values. This hypothesis was tested by collecting consumer-centred data through a questionnaire and analysing and reflecting these findings against a real business case. Even though these data and the case were from Finland, they can be interpreted as representing the

K. Niinimäki (✉)
Aalto University, Espoo, Finland
e-mail: kirsi.niinimaki@aalto.fi

Scandinavian worldview, that is, the lifestyles and societal value bases in Sweden, Finland, Denmark, Norway, and Iceland.

The business case is a well-known design house and Scandinavian lifestyle brand, which tests plant-based natural dyes in their industrial textile printing processes and their garment collections. Natural dyes are not currently largely used in industrial textile colouring processes and are substituted by chemical dyes. Chemical dyes originate from petroleum production and many of them are very harmful and toxic and have environmental impacts. An interest in finding alternatives for synthetic chemicals and colours used in industry is emerging, and natural dyes are gaining attention (e.g., BioColour, 2019).

The research approach presented in this chapter discusses sustainable eco-luxury in a Scandinavian context through the viewpoints of industry, consumers, and the environment.

3.2 Literature Review

3.2.1 Scandinavian Lifestyle

In Scandinavian countries, nature is highly appreciated and understood to be part of the Scandinavian lifestyle and worldview. In Finland, every man's right (*the right to common access, or the right to roam*) provides access to forest areas owned by the state or even by private individuals. This right is significant for Finns and is considered one of the *basic rights* in Finland (Oittinen & Vuolle, 1994). Finns enjoy the right to freely visit forests and to hike, camp, fish, or pick berries in nature. This creates a special relationship with nature and makes recreation possible in natural environments (ibid.). A 2021 study shows that 37% of Finns go out into nature several times a week, 12% every day and 16% once a month or less (Koistinen et al., 2021). Many also traditionally own a family summer cottage and spend their leisure time close to lakes or the seashore, which makes it easy for them to spend time in nature. This tradition also creates a solid foundation for relationships with nature based on positive childhood experiences at summer cottage (Salmi et al., 2006). The roots of many Finns or their parents are still in the countryside (Granberg, 1999), and 'summer cottages provide a way to maintain a connection with the countryside and the peasant way of life' (Salmi et al., 2006, 276). A little over half of the Finnish population spend their summer holidays and weekends in summer cottages close to nature (Sievänen, 2001). It is estimated that the number of leisure residences in Finland is probably the largest in the world relative to population size (Jokinen, 2002).

This special relationship with nature also shows in Finns' mindset. The Finnish population takes environmental protection and climate change seriously. A survey by Eurobarometer in 2013 showed that 84% of the Finnish population considered climate change a serious problem; 57% stated that they had personally taken actions to lower their impact on the environment. The study by Koistinen et al. (2021) supports these earlier findings. The aim of the study was to survey Finns' relationship with nature: 82% of its respondents agreed with the statement that the value of nature cannot be measured in monetary terms, 78% believed that economic and personal wellbeing is dependent on nature, and 71% considered protecting the diversity of nature to be society's most important task. The highest motivators for actions to protect nature were safeguarding future generations' lives (52%), protecting other species and diversity (48%), the desire to act and live according to planetary boundaries (44%), and human beings' dependence on nature (43%). Only 28% of the respondents mentioned their own experiences of the deterioration of nature's conditions as motivation to protect nature (Koistinen et al., 2021).

The same study (Koistinen et al., 2021) revealed Finns' mentality of nature appreciation; most of the participants highlighted how spending time in nature offers intangible wellbeing: how they become calmer, relax, recover from stress, and their mental wellbeing improves. The most important things that Finns felt nature gives them were peace of mind (65%); recreation and energy (62%); mental wellbeing (59%); recovery (56%); access to berries, mushrooms, and fish (56%); and health and physical wellbeing (55%).

Our worldview has a connection to our mentality, which is defined as the unconscious side of worldview (Hyrkkänen, 2002, p. 110). Worldview can also be seen to include notions of human nature, conceptions of the ego, images of society and social stratification, notions of work, perceptions of time and space, and images of a 'good life' (Löfgren, 1981, pp. 26–27). It is connected to our lifestyles, values, and the ways in which we consume. Cultural context and its general worldview define not only the consumption habits of a certain geographical location but also the values behind consumption that lead to certain ways of consuming in general. We also know that environment-related actions such as reducing consumption are affected by a person's own values as well as society's values (Barr, 2003). Lifestyle, as a person's social practices and the story they tell about them, forms the foundations for individuals' consumption choices and the dilemma that they constantly face when making these choices on the basis of their individual needs/desires and social benefits (Niinimäki, 2011; Jackson, 2008). A consumer's sense of self-identify and the understanding of what kind of consumer they are influences their consumption habits (Peattie, 2010). Moral considerations arise when a consumer considers ethical choices and acts in a morally correct way (making the right consumption decision); they approach the ideal ethical world (Oksanen, 2002).

The special relationship with nature can also be seen in consumers' value bases and their consumption habits. So, what kind of consumers are Finns? Finnish consumers are ready to take action to safeguard nature (Koistinen et al., 2021). A clear majority, 62%, was ready to change their lifestyle or consumption habits to stop diversity loss, and 62% were also ready to do the same to stop climate change (Koistinen et al., 2021).

Finnish consumers seem to be slightly less affected by social pressure in their consumption; they are quite individual and base their consumption habits firmly on their own values instead of on social acceptance. In Koistinen et al.'s research, only 24% of respondents stated that they felt pressured to be 'a certain type' or 'to look a certain way,' 27% claimed to be inspired by other peoples' styles or lifestyles, and 31% admitted being influenced by social media, blogs, or magazines (Koistinen et al., 2021). In the same study, the participants pointed out that they only bought things out of real need (64%), but some felt that the need for 'small pleasures' in everyday life might easily lead to impulse buying (60%) (ibid.).

In terms of the products we purchase, especially in the case of textile and fashion items, we need to study ethical consumption. In one study conducted in Finland (Niinimäki, 2011, p. 131), 63% of consumers stated that they were always interested in ethical issues when they were purchasing in general and 28% stated that they often considered ethical issues while purchasing products. When asked about their ethical interest in textiles and garments and how often they considered sustainability when choosing to purchase these products, 17% of consumers responded always and 49% often. These figures show Finns' high interest in ethical and sustainable consumption, and this can be connected to concerns about climate change and the environmental worldview.

3.2.2 New Understanding of Luxury

The traditional understanding of luxury has been that it can be attained by only a few very rich people: elite consumption (Berry, 1994; Appardurai, 1988). However, luxury can be divided into four different categories: utilitarian products and services, indulgence products and services, lifestyle products, and dream luxury (Danziger, 2005). The dream luxury category in particular is connected to very expensive status products through which consumers seek differentiation. Lately we have seen some sort of change in this understanding of luxury. Luxury products have become more common (production has increased) or are largely copied so that brands can also be attained by others and not just the elite few and very rich people (Nyrhinen & Wilska, 2012). Producing and even copying luxury brand products has become so common that they are no longer a luxury, and are more like premium-quality products. Furthermore, rapid technological development and the move towards virtual reality while simultaneously losing touch with nature have increased consumers' needs for authenticity (ibid.). This development is also connected to the emergence of the experience economy, in which people consume more experiences than material products (Pine & Gilmore, 1999). The post-materialistic consumption approach in developed countries (Global North) addresses minimalist consumption, and this can also be seen in the understanding of luxury. Martinez-Allier (1995) connects the new, more responsible understanding of luxury to this new era of post-industrialism and post-materialism.

Nyrhinen and Wilska (2012) have studied the change in the understanding of luxury among Finnish consumers. In addition to luxury as a status orientation, a new definition has been emerging: it includes authenticity, experiences, and ethical and ecological considerations. These findings clearly show that Finnish consumers respect more ethical and ecological aspects of luxury than the status aspect. Even though luxury is still linked to a hedonistic value base of consumption, the results from Finland can be understood as part of a greater international development in which consumers' interests are shifting to a greater extent than before from materialist product-owning and status-showing to more abstract luxury understanding and an immaterial approach to luxury, which places more importance on company responsibility (ibid.).

A study conducted in Finland shows that even general consumption trends are changing from materialistic interest towards a more unmaterialistic way of consuming (Statistic central, sited by Niskakangas, 12.8.2019). In the mid-1990s, Finnish consumers spent as much money on services as they did on products, but since then the money spent on services has increased steadily. For example, in early 2019 during the first quarter, money spend to services was 26% more than money spent on products (ibid.). The consumption trend has been to simplify home interiors (minimalism and the KonMari movement) and decrease the number of products one owns and instead to invest in one's own wellbeing, experiences, and services. This also seems to differ between generations: the older generation respects the material side of consumption (e.g., products with sentimental value) while the younger generation wants to own as few objects as possible and, instead of buying new products, to invest in experiences. The notion is that one can increase their happiness with services and even achieve longer-lasting satisfaction than from buying new products (ibid.).

3.3 Survey on Consumers' Understanding of Luxury

A consumer survey was conducted in spring 2019 (March–April) in Finland on consumers' understanding of luxury in general and its connection to the textile and garment field, food sector, travel, and service experiences. The survey link was distributed on social media platforms (Twitter, Facebook, some discussion groups), and was based on the snowball sampling approach. The survey reached 94 respondents, of whom 10 were male and 84 were female consumers. Seventy-four respondents informed us that they lived in a city and 20 informed us that they lived in the countryside.

The questionnaire had both statement types of questions (based on a seven-point Likert scale) on the different aspects of luxury, structured multiple choice questions and open questions, which the respondent could answer in their own words. For example, 'What does luxury mean to you? Is it material, experiences, emotions, own time, all of these, or something else? Please describe it in your own words.'

The mixed method approach was applied to develop a more comprehensive and descriptive understanding of the phenomenon under study (Patton, 1999). The analysis was conducted in three phases. In the first round, the answers that were countable were listed according to their numerical value (e.g., background information) to build a picture of the respondents' age groups and income levels as well as their general understanding of luxury. Tables 3.1 and 3.2 show the respondents' age groups and income levels. Table 3.3 shows the consumers' eagerness to seek luxury experiences from food, fashion, services, and travel, as well as the frequency of such purchases. Table 3.4 shows the respondents' emerging positive feelings towards garments' different fibre and colour types. To further survey consumers' luxury experiences, multiple choice questions

Table 3.1 Respondents' age

Respondents' age n=						
Below 20 years	21–30	31–40	41–50	51–60	61–70	Above 70
4	8	10	23	25	22	4

Table 3.2 Respondents' income levels

Respondents' income level n=				
Low income	Lower middle class	Middle class	Upper middle class	Upper class
25	17	22	17	13

Table 3.3 Sources of luxury and frequency of searching for it

	Cannot say n=	Never n=	Seldom than once a year n=	A few times a year n=	Once a month n=	Weekly n=
Food	1	4	11	22	34	22
Fashion	1	11	19	40	11	11
Services	3	7	18	36	26	4
Travel	3	8	23	41	6	3

Table 3.4 Positive image of colour and fibre sources (natural or synthetic)

Positive image in connection to	Natural textile colours n=	Synthetic textile colours n=	Natural fibres n=	Synthetic fibres n=
Environmental responsibility and sustainability	72	4	72	4
Recyclability	58	4	78	17
Luxury	64	6	63	1

were used, which offered a more detailed picture of all the elements that luxury experiences could offer in the context of garments. Table 3.5 illustrates these results.

The second stage examined the open answers using the content analysis approach as the guiding principle to find common themes related to luxury and the answers to the research question on how Finnish consumers understand and define luxury. The analysis included several rounds to account for rigour and to form a solid understanding of the key themes of this topic (Flick, 2004). These findings are expanded and described in more detailed in the 'Three categories of luxury' section. Table 3.6 summarises the findings of this analysis. The third stage involved the

Table 3.5 What elements affect consumers' experience of luxury (in clothing)?

Garment and luxury	Total n = 94
A garment being of high quality and long lasting.	n = 88
A garment feeling comfortable when worn.	n = 77
Knowing that the garment's manufacture has required a great deal of handcraft work.	n = 74
The company behind the garment transparently revealing their supply chain and manufacturing process.	n = 73
Feeling that the garment is made for me.	n = 72
The garment being easy to maintain (e.g., water laundering).	n = 72
The garment being made locally (e.g., in Scandinavian countries).	n = 66
The garment being made in Finland.	n = 64
The garment's manufacture having a low environmental impact.	n = 64
The garment being made from natural materials.	n = 64
The company behind the garment being responsible (social responsibility, environmental responsibility).	n = 64
The garment being coloured using natural colours.	n = 61
The garment being from a famous brand.	n = 32
The garment being designed by a famous designer.	n = 26

interpretation of the direction of the luxury experience, that is, the direction that the respondents mentioned when describing what luxury means to them. They claimed that some aspects of luxury pointed towards looking for social acceptance (outwards), rewarding yourself (inwards) and even towards personal wellbeing (inwards). These directions are shown in Table 3.6. To deepen the description of this phenomenon in this special context, the Marimekko case was used as an example of the value base in Scandinavian eco-luxury.

3.4 Results

The main finding in these answers was that Finnish consumers define three categories of luxury; (1) an elite lifestyle (only a few people able to attain it), (2) products or services that are one grade better (requires some effort to attain it), (3) everyday luxury (i.e., services that cheer you up) (see Table 3.6). These survey results will be discussed in more detail in

Table 3.6 Finnish consumers understand luxury through three different categories

	Lifestyle	One grade better	Everyday luxury
Quotations	'Something that exists in only a few pieces, which are especially good, premium quality.' 'Products that are made especially expensive and not needed in everyday life.'	'It's the kind of thing that hurts a bit to get, but I want to reward myself.' 'Something you don't necessarily need but gives you good vibes.'	'Everything that is above your everyday experience.' 'That lifts you up from your everyday life.' 'A cup of coffee, from a small roastery and good service in a small café.' 'Experiences in nature.' 'Moments with friends.'
Attributes	Not easy to attain Only for a few consumers Overall lifestyle, including materialistic consumption Negative luxury=only for the elite, unnecessary products	Product or service Attaining it requires some effort Worth all the effort Monetary investment Indulging yourself	Experience, moments Connected to everyday products or immaterial services Makes your day slightly better Adds value to your day Free time, wellbeing
Attributes connected to products	Aesthetic products, haute couture Premium-quality brand products International-level high fashion Top-quality materials: silk, merino wool, gold	Good materials Eco-products and eco-materials Quality Durable, long-lasting Special, unique Timeless design Transparency in supply chain	Made in Finland Easy to take care of (e.g., water laundry)
Direction	Outwards: Looking for social acceptance	Inwards: Rewarding yourself	Inwards: One's own wellbeing

Sect. 3.4.4, but first we present the sources of luxury that people search for, the eco-aspects of luxury, and luxury experiences of garments.

3.4.1 Sources of Luxury

When asked about the sources from which Finnish consumers searched for their experiences of luxury, they answered fashion, services, and travels as the areas to which they turned a few times a year. Food was a field they searched for the luxury experience once a month, but some consumers used food as a luxury experience even weekly (see Table 3.3).

3.4.2 Eco-aspects of Luxury

As one aim of the study was to identify consumer's attitudes towards natural colours and natural fibres, it asked: what kind of image arises in your mind when you think of these aspects (positive or negative image)?

Table 3.4 shows that in consumers' minds, the image of natural textile colours and natural fibres in textiles and garments had a strong connection not only to sustainability but also to the image of luxury. The connection to synthetic colours or synthetic fibres in textiles and garments was not so strong.

3.4.3 Experience of Luxury in Garments

The questionnaire also asked respondents' opinions on what kinds of elements in clothing affect their experience of luxury. High quality and durability received the highest number of responses but wear experience was also highly appreciated. Several answers connected to sustainability and responsibility also mentioned transparency in the supply chain, location of manufacture, environmental impact, and natural materials (see Table 3.5). It can be interpreted that consumers connect luxury not only with better quality but also with deeper environmental consideration and greater responsibility more than do other product types. This is an interesting finding, as these aspects are less obvious in the current global fashion business in which manufacturing factories are often on the other side of the globe, transparency is hard to achieve (supply chains are long), and the control of sustainability issues or environmental impacts is very difficult.

On the other hand, well-known brands or famous designers received a low number of responses, which can be interpreted as the status aspect of luxury not being very important for these consumers.

3.4.4 Three Categories of Luxury

Analysis of the open responses to what luxury means for consumers revealed three categories: elite lifestyle, one grade better, and everyday luxury (see Table 3.6).

Elite Lifestyle
This lifestyle category represents the traditional way of understanding luxury. Comments such as luxury is 'something of which only a few pieces exist but are of especially good, premium quality' and 'products that are made especially expensive and are not essential in everyday life' point out the aspect of being unnecessary, rare, and only for a few people. This category was also linked to a strong materialistic way of consuming.

From the product design point of view, consumers highlighted attributes such as a high level of aesthetics and quality, a well-known brand or designer, haute couture in fashion, and top-quality materials. Durable products were also mentioned, as well as expectations that luxury products are made with more consideration of ethical aspects in production and more transparency in the supply chain. The interpretations of this is that Finnish consumers expect more responsibility and sustainability from luxury products and that this is morally and ethically the correct way in which to act. Moreover, knowledge of the background of the product, transparency in general, and environmental impacts, in other words not only responsibility but also transparency and more information, were connected to luxury products.

A material luxury is when I know exactly what it is (origin, production process, environmental impacts), how I should take care of it and how it lasts—it has to be long lasting. Nothing disposable represents luxury to me.

Furthermore, the aspect of differentiating these brand products was mentioned.

When I buy some luxury garment, I expect no one in my neighbourhood to have the same garment.

Consumers build their identity in a social context (Kaiser, 1990) and this elite lifestyle category can be seen to link the aspect of trying to reach social acceptance through identity-building with external symbols such as brand garments. McCracken (1988) pointed out that products can be seen to represent a certain lifestyle that consumers are trying to reach. Brand products can be understood as building a bridge towards the desired lifestyle and the existing reality. Consumers purchase brand garments to obtain a small part of the desired lifestyle to which they aspire (ibid.). This could be connected to the dream aspect, when a luxury product is something we dream of owning, but we also dream of having all the values (status) that are connected to this brand and the lifestyle it represents.

One Grade Better

The second category that could be identified from the answers was *one grade better*. Attaining this requires some effort and/or monetary investment. 'It is the kind of thing that hurts a bit to get, but I want to reward myself.' Reward or indulgence was mentioned often. One grade better requires effort, an extra push, or endeavour. This it is not so easy; it requires more time and therefore moves the fulfilment of expectations from immediate gratification. This aspect of waiting and dreaming are unexceptional in these times of fast fashion purchasing and easy fulfilment, as this fulfilment happens especially during impulse fashion shopping (Niinimäki, 2018).

This category is linked to rewarding oneself and raising oneself slightly above everyday consumption. 'Something you don't necessarily need, but gives you good vibes.' This could be linked to emotional consumption needs and the meaning of garments for building one's identity; not in a social context in this category, but more on an individual, private level.

> For me luxury is a quality that can even exist in small details. For example, in a garment it can be good material or a beautiful cut.

> Luxury for me is a life that looks and feels like me. Little but good. I search for beautiful and long-lasting, natural materials in my life, and I enjoy my everyday life. I don't spend, but I do get things I really need: functional, high-quality products, which I really enjoy and are long lasting and durable.

Several answers could be connected to product attributes such as durability and long-lasting, timeless styles: unique design and quality materials. Sustainability elements could also be identified such as transparency in the supply chain and eco-materials. These show that a sustainable product is understood to be 'a bit better' than the average product. In a time when fast fashion is a dominant product type in markets and is connected to low-quality products and unsustainability, recognising this is important. Consumers understand eco-products and sustainability to be 'a bit better,' something that requires more effort, and could offer a better, more sustainable value base as well as better product quality.

Everyday Luxury

> First, I thought elegance and high quality, but when I thought more carefully, it could be something very simple, which I enjoy myself.

The abstract level was visible in the third category of luxury, which can be defined as *everyday luxury*. Everyday luxury contained more experiences than a physical product. It could be linked to a certain moment that 'lifts you up,' 'makes your day a bit better,' or adds 'value to your day.' Very often, these passing moments were connected to a pause, free time, and one's own wellbeing. Experiences with nature were mentioned several times, which connects them to environmental values and worldview and to the Finnish way of living close to nature.

These things do change. Luxury is moments when I can be all alone, in total peace and quiet, I can go to the sauna to relax, hike in a forest and generally be in nature.

Experiences that don't have to be expensive but create unforgettable memories.

New experiences linked to learning a new skill, starting a new hobby or services connected to wellbeing were also mentioned.

For me, luxury is my own time, my knitting, beautiful knitting yarns, learning a new skill on a course.

Social moments, such as meeting a friend in a café, a family dinner, or an evening out with a friend were also mentioned.

An evening at the theatre, which includes dressing up and dinner.

I don't want luxury products, but good experiences in my life.

Finnish consumers respect functionality and easiness in everyday life. Examples of elements of everyday luxury linked to products were attributes that make everyday life somewhat easier, such as easy garment maintenance, but also textiles and garments made in Finland (valuing local production).

Understanding one's own wellbeing shifts the focus in this category strongly inwards, adding pleasure to one's own everyday living, cheering oneself up, and finding a better balance in life. This may be connected to the Scandinavian worldview and slower way of living, balancing experiences connected to nature, or social moments with family or friends. The aspect of finding happiness and longer-lasting satisfaction through immaterial experiences was strongly present in this category.

3.5 Marimekko Case

Marimekko is an iconic Finnish company that represents the Scandinavian lifestyle, and whose production is based on clear design, simple cotton fabric, and bold prints. Marimekko looks for sustainable alternatives suitable for their industrial manufacturing practices. Recently, Marimekko has been testing plant-based natural dyes, which are not so easy to use in industrial manufacturing. Most of the colours used in industrial fabric printing are synthetic, with origins from non-renewable sources (mostly from petroleum production). Many harmful chemicals are used in industrial dyeing and printing processes. For example, synthetic indigo, originating from oil production, is produced through high-energy processes, and even cyanide and formaldehyde are used in its production process.

The blue colour originating from natural woad plant (*Isatis tinctoria*), which Marimekko has used in its 2021 autumn collection, was cultivated in Finland. Its origin can even be identified on the field level and in this way its production is transparent and can be defined as super local. A new kind of relationship with the colour plant farmer is established, which is quite exceptional in the larger-scale textile industry. Testing this plant-based dye on an industrial scale has required a great deal of experimentation and negotiation with a farmer to ensure that the dye is suitable for industrial processes and that the end result is successful and durable in Marimekko's processes.

By selecting more sustainable production alternatives, Marimekko aims to reduce its environmental impacts. Natural colours come from renewable sources, and environmental issues such as carbon sequestration can be considered in their cultivation process (Ammayappan & Jose, 2015). Plant roots can sequester carbon, and this activity can be maximised through different agricultural practices such as crop rotation, selecting plants with deep roots, and using ground cover plants. All these aspects can be realised in woad cultivation. As the availability of this colour source (plant-based colour) is limited and production (harvesting) only takes place once a year, it will end up in a limited edition collection and result in a new rhythm in industrial manufacturing and in a slower eco-design approach, which is an alternative to fast fashion and its current industrial rhythm.

The designs chosen for this capsulate collection are the iconic Marimekko prints Stripe (Piccolo) and Flower (Unikko) which every Marimekko fan easily recognises. In this way, Marimekko can combine its design value and brand value with environmental values in this capsulate collection. Using natural dyes in its production and providing an alternative approach for industrial garment manufacturing enables Marimekko to implement the following values in its product design; super-locality, transparency, natural processes, ecological production, uniqueness, and iconic style.

The image and identity of Finnish design consists of functional, nature inspired, natural-based material choices; clean, simple, and modern in its aesthetic language (visual presentation and forms) (Savolainen, 2008). Marimekko represents all these values in its design. As Finnish design has an ethos of authenticity, uniqueness, but also a yearning to find its own roots, it has a strong connection to the Finnish value base and the Finnish worldview (ibid.). All these aspects are strongly represented in Marimekko, its designs and its cultural meaning. The Finnish blue colour, in combination with the iconic Marimekko print design, and the simple functional garment design of cotton fabric communicates the message to consumers that although it is eco-design, it is a slightly more exceptional, limited edition; more authentic and more sustainable. Super-locality and transparency add nicely to Marimekko's narrative of being a Finnish sustainable design brand. As these aspects are quite rare in the global fashion market, they give a certain 'eco-glitter' to this collection. In this way, Marimekko can be seen as representing eco-luxury, which belongs to the one grade better category—slightly more exceptional but still attainable. Everyday luxury aspects are also offered by the functionality and easy-care aspects. This kind of product gives consumers satisfaction on two levels: first by providing a functional and aesthetic product, and second by offering ideological satisfaction through a sustainable value base.

3.6 Conclusions

One interesting finding of the study was consumers' expectations that luxury garments are more sustainable, production is more transparent and that the companies behind luxury products are more responsible than the average fashion company. This finding provides a new understanding of luxury. Most previous studies have shown that consumers do not think about ethical or environmental issues when considering luxury products. Davies et al. (2012, p. 46) highlight that the ethical issues of luxury goods do not interest consumers as much as the ethical considerations connected to other goods. Streit and Davies (2013, 209) point out that 'if luxury good consumption is about hedonism and self-pleasures … it is less likely that altruism would override hedonistic desires.' Yet the findings of the current study in Finland show that a new understanding of luxury might be emerging. The responsibility of companies could be seen in the open answers: the respondents expected luxury to mean more sustainable and more ethical, and companies to be more transparent in their actions. In Streit and Davies' (2013) study, consumers expected to find more information about these issues, whereas in the current study, consumers expected this information to already exist and be available.

In general, the consumers were not so interested in luxury, at least if we define luxury as highly expensive products and an elite lifestyle. The new notion of sustainable eco-luxury with a new kind of ethical consideration and company responsibility might provide a better and more suitable category of luxury products which are easier for larger groups of consumers to accept. Streit and Davies (2013) found the same in their study. They documented a lack of fundamental interest in ethical luxury fashion. Consumers were more interested in quality, materials, and timeless design than the aspect of luxury (ibid.).

To succeed in luxury markets, products or services need to fulfil consumers' expectations. As Davies et al. (2012, p. 14) highlight 'for ethical-luxury to work, it would therefore need to enhance (or at least not destroy) these self-pleasure and hedonic aspects of luxury consumption.' These self-pleasures and hedonistic aspects are tied to cultural context. Accordingly, it is important to understand the market segment, geographical location, and what this framing brings to consumers' worldviews and lifestyles and how this value base affects consumers' luxury understanding and the elements that consumers look for in luxury—not only the self-pleasure and hedonistic aspects of luxury consumption, but also the fundamental values in their worldview in particular context.

Based on this small study, it can be stated that in the Scandinavian context, luxury is a well-functioning, high-quality, and longer-lasting aesthetic product or service that offers wellbeing or pleasure. Moreover, luxury products and experiences can also be based on Scandinavian consumption values and the special relationship with nature. Eco-luxury in the Scandinavian context also includes the company's responsibility and in the best case, local production, and more transparent production, as can been noted in the Marimekko case. The eco-aspect in luxury means a smaller environmental impact and slower design and manufacturing processes. Moreover, the aspect of slowness means more consideration and less production in total (more accurate production, capsulate collections, limited editions). Sustainability in this context can mean transparency, more local production, and even a new kind of relationship with the farmers who produce colour plants. Scandinavian design 'language' presents clear, long-lasting aesthetics, and a well-functioning product that is easy to wear and maintain.

The slow approach in industrial production also provides an alternative to current fast mass-manufacturing and might positively differentiate sustainable eco-luxury in consumers' minds. As Strauss (2015, p. 88) highlighted while studying the aspect of slow production in the fashion field, it might require 'recalibrating of the way they think, reconsidering how they've grown accustomed to getting things done and re-examining inherited belief systems and practices at the very heart of their chosen fields.' This points out that sustainable eco-luxury may not only be a way of focusing on and implementing the environmental value base but also be an option to do things differently. Production and even its scale can be considered more, and even a new kind of partnering can be constructed in the local context. This aspect enables a real alternative to be constructed as a sustainable eco-luxury.

As pointed out by Karaosman et al. (2020), when a company is developing a proactive attitude towards sustainability, especially in a complex business environment such as fashion, it can create a competitive advantage. The unique characteristics of a product or production can create a pioneering role for a company, as it benefits from being a 'first mover' and doing something differently to its competitors (ibid.). Moreover, cooperation in the supply chain is fundamentally important, especially when limited resources are in question (Govindan, 2018). As the Marimekko case shows, and as supported by Karaosman et al. (2020), knowledge-sharing and joint learning among different partners are essential when new sustainable practices are under development.

Although the value of luxury markets is decreasing due to, for example, large-scale copying, this study highlights possible options for luxury brands. It is not only high quality and monetary value that are associated with luxury items; authenticity and true narrative are also now more connected than ever to luxury brands and luxury products. These, as well as sustainability, are becoming increasingly important aspects for consumers, even in the luxury markets. Locality and cultural context should be taken into account and seen as positive references for designing unique products and generating small-scale production (e.g., capsulate collections) that could include a nonmaterialist value base, for example, ecological or sustainable values. Consumers also eagerly seek unique experiences more than unique products, in connection to their sustainability worldview. Therefore luxury brands could provide immaterial experiences more than material ones. Moreover, consumers expect luxury brands to be more sustainable and more transparent in their actions than average companies, which forms a new sustainable value basis for luxury brands.

Research Findings

- Consumers' understanding of luxury is changing and is moving away from status products to services.
- Consumers' understanding of luxury has the following directions: outwards (looking for social acceptance and differentiation), inwards (rewarding oneself), and inwards (enhancing one's own wellbeing).
- Consumers expect greater responsibility, transparency, and sustainability from luxury products and the companies behind them than from other product categories.

Managerial Implications

- Focusing on sustainable eco-luxury provides alternative ways of designing and producing and even creating a slower rhythm or smaller scale in industrial production.
- It is important to take into consideration the geographical location, cultural context, and consumers' worldview and lifestyle to be able to

produce products whose value bases are suitable for consumers in specific locations.
– Consumers respect authenticity and a true narrative behind a product and this can be enhanced through transparency.

Acknowledgements This research was supported by the Academy of Finland's Strategic Research Council, grant number 327 330 Bio Based Dyes and Pigments for Colour Palette (BioColour).

References

Ammayappan, L., & Jose, S. (2015). Functional aspects, ecotesting and environmental impact of natural dyes. In S. S. Muthu (Ed.), *Handbook of sustainable apparel production* (pp. 333–350). CRC Press.
Appardurai, A. (1988). Introduction: Commodities and the politics of value. In A. Appardurai (Ed.), *The social life of things, commodities in cultural perspective* (pp. 3–63). Cambridge University Press.
Barr, S. (2003). Strategies for sustainability: Citizens and responsible environmental behaviour. *Area, 35*, 227–240.
Berry, C. (1994). *The idea of luxury. A conceptual and historical investigation.* Cambridge University Press.
BioColour. (2019). *BioColour – Bio-based dyes and pigments for colour palette.* https://biocolour.fi/
Danziger, P. M. (2005). *Let them eat cake: Marketing luxury to the masses – As well as the classes.* Dearborn Trading Publishing.
Davies, I. A., Lee, Z., & Ahonkhai, I. (2012). Do consumers care about ethical-luxury? *Journal of Business Ethics, 106*, 37–51.
Eurobarometer. (2013). *Climate change*, Spec. Eurobarom., 409.
Flick, U. (2004). Triangulation in qualitative research. In *A companion to qualitative research* (Vol. 3, pp. 178–183). Sage.
Govindan, K. (2018). Sustainable consumption and production in the food supply chain: A conceptual framework. *International Journal of Production Economics, 195*, 419–431.
Granberg, L. (1999). Introduction. *Sociologia Ruralis, 39*, 277–279.
Hyrkkänen, M. (2002). *Aatehistorian mieli* [The mind of the history of ideologies]. Tampere.

Jackson, T. (2008). The challenge of sustainable lifestyles. In *State of the world, innovations for the sustainable economy* (pp. 45–60). The Worldwatch Institute, W.W. Norton & Company.

Jokinen, A. (2002). Free-time habitation and layers of ecological history at a southern Finnish lake. *Landscape and Urban Planning, 61*, 99–112.

Kaiser, S. (1990). *The social psychology of clothing. Symbolic appearances in context* (2nd ed.). Macmillan.

Karaosman, H., Perry, P., Brun, A., & Morales-Alonso, G. (2020). Behind the runway: Extending sustainability in luxury fashion supply chains. *Journal of Business Research, 117*, 652–663.

Koistinen, A., Lehtinen, A, & Nieminen, E. (2021). *Suomalaisten luontosuhteet* [Finnish population's relationships with nature]. Sitra & Kantar TRS. https://www.sitra.fi/julkaisut/suomalaisten-luontosuhteet-kysely/

Löfgren, O. (1981). World-views: A research perspective. *Ethnologia Scandinavica, 11*, 21–36.

Martinez-Allier, J. (1995). The environment as a luxury good or "too poor to be green"? *Ecological Economics, 13*(1), 1–10.

McCracken, G. (1988). *Culture and consumption. A new approach to the symbolic character of consumer goods and activities.* Indiana University.

Niinimäki, K. (2011). *From disposable to sustainable. The complex interplay between design and consumption of textiles and clothing.* Doctoral dissertation. Aalto University.

Niinimäki, K. (2018). Knowing better, but behaving emotionally: Strong emotional undertones in fast fashion consumption. In C. Becker-Leifhold & M. Heuer (Eds.), *Eco friendly and fair: Fast fashion and consumer behavior* (pp. 49–57). Routledge.

Niskakangas, T. (12.8.2019) Vähemmän Tavaraa, Enemmän Palveluita [Less products, more services], Helsinki News Paper, A24-A25.

Nyrhinen, J., & Wilska, T. A. (2012). Kohti Vastuullista Ylellisyyttä? Eettiset ja Ekologiset Trendit sekä Luksuskulutus Suomessa [Towards responsible luxury? Ethical and ecological trends and luxury consumption in Finland]. *Kulutustutkimus.nyt, 6*(1), 20–41.

Oittinen, A., & Vuolle, P. (1994). 'Jokamiehenoikeus: perinteistä nykypäivään' [A research project: The right of common access]. *Liikunnan ja Kansanterveyden julkaisuja, 92*, 53.

Oksanen, R. (2002). *Suomalaisten Kuluttajien Suhtautuminen Eettiseen Kaupankäyntiin ja Reilunkaupan Tuotteisiin* [Finnish consumers' attitudes towards ethical trade and fairtrade products]. Jyväskylä University.

Ollila, A. (1998). Perspectives to Finnish identity. *Scandinavian Journal of History, 23*(3–4), 127–137.

Patton, M. Q. (1999). Enhancing the quality and credibility of qualitative analysis. *Health Services, 34*(5 Pt 2), 1189–1208.

Peattie, K. (2010). Green consumption: Behavior and norms. *The Annual Review of Environment and Resources, 35*, 195–228. https://doi.org/10.1146/annurev-environ-032609-094328

Pine, J. B., & Gilmore, J. H. (1999). *The experience economy*. Harvard Business School Press.

Salmi, P., Toivonen, A.-L., & Mikkola, J. (2006). Impact of summer cottage residence on recreational fishing participation in Finland. *Fisheries Management and Ecology, 13*, 275–283.

Savolainen, J. (2008). Kulta-ajasta Kulta-aikaan [From golden-time to golden-time]. In M. Aav, J. Savolainen, & L. Svinhufvud (Eds.), *Fennofolk: New nordic oddity* (pp. 13–17). Design Museum.

Sievänen, T. (Ed.) (2001). *Luonnon Virkistyskäyttö 2000* [Outdoor Recreation 2000]. Metsäntutkimuslaitoksen tiedonantoja (Vol. 802).

Strauss, C. (2015). Speed. In M. A. Gardetti & A. L. Torres (Eds.), *Sustainability in fashion and textiles* (pp. 82–90). Sheffield.

Streit, C., & Davies, I. (2013). Sustainability isn't sexy: An exploratory study into luxury fashion. In M. A. Gardetti & A. L. Torres (Eds.), *Sustainability in fashion and textiles* (pp. 207–222). Greenleaf.

4

Sustainable Luxury: A Framework for Meaning Through Value Congruence

Stephanie Y. Volcon, Marian Makkar, Diane M. Martin, and Francis Farrelly

4.1 Introduction: The Imperative of Sustainability

The word "sustainability" has firmly secured a home in the contemporary vernacular. Sustainability originated from forestry as the "silvicultural principle that the amount of wood harvested should not exceed the volume that grows again" (Carlowitz 1713 quoted in Geissdoerfer et al., 2017: 758). A more contemporary meaning of sustainability is connected to global warming, aka climate change (Mulkerrins, 2021). The United Nations report on climate change (IPCC, 2021) notes that that the maintenance of the current level of greenhouse gas emission creates a high probability of a major climate catastrophe in the near future. In order to prevent this highly likely scenario, an urgent decrease of human induced greenhouse gas emissions will present a discernible difference in the global surface temperature within the next 20 years (ibid.). Mittelstaedt

S. Y. Volcon (✉) • M. Makkar • D. M. Martin • F. Farrelly
Marketing Department, RMIT University, Melbourne, VIC, Australia
e-mail: s3817303@student.rmit.edu.au

© The Author(s), under exclusive license to Springer Nature Switzerland AG 2022
C. E. Henninger, N. K. Athwal (eds.), *Sustainable Luxury*, Palgrave Advances in Luxury, https://doi.org/10.1007/978-3-031-06928-4_4

et al. (2014) argue that sustainability is a megatrend, one that initiates a seismic shift in time, either highly visible in its effect, such as printing press invention, or so omnipresent and contextual, that this shift is often taken for granted. While climate deniers (Norgaard, 2011) may refute the megatrend moniker, the irreversible effect of climate change now leads to the inclusion of sustainability in most global and national institutional discussions, business agendas and mainstream consumer discourses.

Concern for a sustainable future influences most aspects of society: political, technological, economic, and cultural. For example, the United Kingdom has committed to an 80% decrease carbon emission by 2050, based on 1990 levels (UK Government, 2008). The United States aims to cut 50% of its current greenhouse gas emissions by 2030 and have a carbon-free power system by 2035, demonstrating a commitment of structural change by the country most responsible for energy use (The White House, 2021). Sustainability has become a major source of technological innovation (Markard et al., 2012). Pension funds have begun to disinvest from extraction industries (Hansen & Pollin, 2020). Even the august body of the World Bank has taken action and released €163 million worth sustainability bonds to support project implementation efforts (Ernst & Young, 2017). Major higher education institutions including Harvard University have pulled out oil-related financial investments (The Guardian, 2021). Eighty-eight per cent of business students now see learning about sustainability related issues as a priority (Hoffman, 2018). In short, governments and major social institutions are now more aligned with sustainable goals that have long been of concern to environmentalists (Carson, 1962), activists (McKibben, 2008) and an ever-increasing number of consumers (Koch, 2021).

Sustainable consumption offers more than just a rational calculation of long-term survival of life on the planet. Many people pursue sustainable behaviour for its "double dividend," which refers to acquiring both the external benefit of helping the environment as well as the personal benefit of improving one's life either practically or emotionally (Jackson, 2005). For instance, boycotting unsustainable brands empowers consumers by demonstrating how they can impact the marketplace (Connolly & Prothero, 2008; Roux & Izberk-Bilgin, 2018; Smith, 1990). Green purchases allow consumers to self-express as active citizens (Cherrier et al., 2011; Connolly & Prothero, 2008), and anti-consumption movements provide people with feelings of happiness and control over their lives (Chatzidakis & Lee, 2013; Connolly & Prothero, 2008; Lee & Ahn, 2016). These findings suggest that sustainable behaviour is driven by a combination of rational, emotional, and symbolic goals and desires.

4.2 Luxury: The Height of Consumption

Consumption of luxury goods is equally multifaceted. Luxury consumption does connote not only financial status but also a rich cultural heritage and symbolic associations (Dubois & Duquesne, 1993; Kapferer, 1997). Emotionality is evident in the luxury field, where the use of gratitude appeals made people more inclined to share the information about sustainable luxury brands with a wider online audience (Septianto et al., 2021). Luxury products occupy the top end of the fashion hierarchy and as such exert substantial influence on the entire industry and society at large (Kapferer, 2010). While researchers have developed theoretical perspectives on sustainability in fashion in general (Fletcher, 2013), we join the conversation on sustainability in the luxury market (Athwal et al., 2019; Keinan et al., 2020; Kunz et al., 2020).

In this conceptual chapter we argue that investigation of congruent values between sustainable and luxury consumption can enhance our understanding of what cultural meaning sustainable luxury holds for consumers and how luxury producers can engage sustainability to enhance the essence of their brands. We provide a typology of both luxury and sustainable values and demonstrate how one luxury brand successfully links these values.

4.3 Luxury and Sustainability: A Contested Partnership

Luxury has long suffered from a compatibility problem when it comes to sustainability. Luxury holds the highest symbolic power of all areas of consumption, and as such is regarded as a benchmark and aspirational trendsetter—for producers, consumers, and within general public (Kapferer, 2010). While other industries moved towards more sustainably responsible production and consumption, luxury dragged its heels. This lack of leadership in sustainable development was widely condemned after the 2007 *Deeper Luxury* report revealed that the ten biggest luxury companies were graded at only 67–69 out of 100 points scale for their social and environmental responsibilities (Bendell & Kleanthous, 2007).

There is some question whether luxury and sustainability are even compatible. Kapferer and Michaut-Denizeau (2014) posit that if luxury stands for exclusivity and appraisal of craftsmanship, and sustainability holds democratic beliefs of saving, maintaining, and sharing environmental resources with others, it may not be possible for the two to cross-pollinate. Some consumers more readily associate luxury with unsustainability than with sustainability (Voyer & Beckham, 2014). Yet others reject this idea, arguing that some consumers see sustainable luxury products more favourably (DiDonato & Jakubiak, 2016), and others find parallels between sustainability and luxury evident in the organic food and real estate industries (Fifita et al., 2020; Fuerst & Shimizu, 2016). Times are changing, and consumers are looking for sustainability in luxury goods (Cervellon & Drylie Carey, 2021).

There is also a lack of clarity about what sustainability means with respect to luxury (Dean, 2018). Since luxury feeds from cultural sources and values of the social elites (Brooks, 2001), sustainable luxury depends more on the historic-cultural context in addition to technical innovation for further integration of a sustainable ethos. Athwal et al.'s (2019) systematic literature review on the topic goes some way towards explaining these difficulties. They found that the research on sustainable luxury is often contradictory and underdeveloped, partially due to the changing meaning of luxury (Athwal et al., 2019). For example, traditional premium-priced goods have not only gone mainstream (Danziger, 2005) but are now consumed by individuals in lower 50–60 percentiles of the population income scale (Currid-Halkett, 2017); more people of modest means are consuming luxury goods. The meaning of luxury is also changing at the other end of the economic scales where traditional higher-class consumers see their frivolous spending as an unearned privilege (Goor et al., 2020) and are capable of redefining the social and cultural understandings of luxury (Banister et al., 2020; Belk, 2020). Some examples of what is now considered luxury include non-materialistic manifestations, such as busyness and the lack of free time (Bellezza et al., 2017), human contact instead of online education (Bowles, 2019), electric cars instead of petrol cars (Ottman et al., 2006), fair trade and artisanal products instead of luxury brands (Brooks, 2001), and curated material consumption, especially among upwardly mobile global youth (Atanasova & Eckhardt, 2021).

The luxury industry is faced with upholding its dream-like aspirational status while maintaining brand equity as it offers relevant products to satisfy the changing tastes of the sustainable luxury consumer. One effort to demonstrate luxury goods as sustainable is the "Butterfly Mark," a certification is awarded by the Positive Luxury company to luxury brands that provide transparency by exhibiting their company's sustainable operations, similar to the food industry's Fair Trade label (Keinan et al., 2020). This certification aims to serve as a visible symbol of the sustainable impact the company makes and to signify its efforts in transparency and building trust with consumers and other stakeholders whilst avoiding greenwashing claims (*Positive Luxury*, 2022). The existing shift towards more sustainable production currently reflects only a small percentage of the luxury industry which has a long way to go in terms of an imperative industrial and cultural transition in the face of the global sustainability megatrend.

4.4 Value Congruence in Luxury and Sustainability

On the surface it seems that luxury and sustainability are diametrically opposed value systems. Indeed, Keinan et al. (2020) find the most commonly used values associated with luxury are pleasure, self-indulgence, superficiality, and ostentation, whereas sustainability is typically linked with sobriety, altruism, moderation, and ethics. However, high quality, excellence, and durability are components equally intrinsic to luxury and sustainable consumption (Athwal et al., 2019; Keinan et al., 2020). Sustainability and luxury also have similar values of timelessness and heritage: whereas luxury brands are associated with timeless use, heritage, and long history, sustainability too aims to preserve existing natural resources from present to future generations for its indefinite existence (Joy et al., 2012). Some fashion brands now turn to slow fashion with the aspiration for timelessness and "forever" fashion items (Aleksander, 2020). Rarity is another common value for both luxury and sustainability discourses: luxury is dependent on exclusive rare resources and sustainability signifies protecting exclusive rare natural resources that are non-replenishable (Kapferer, 1997). The notions of beauty are equally important for luxury brands that source their inspiration from art; the notion of sustainability and sustainable ideologies strive to save natural beauty from waste and overconsumption (Joy et al., 2012; Kapferer, 2010). See Table 4.1 for a summary of common values intrinsic to both luxury and sustainability.

Kapferer (2010) proposes that the representation of a sustainable lifestyle as the way to stop climate change and enjoy a cleaner and safer planet could qualify for a new "dream-for" aspiration as the essence of luxury as an ephemeral dream. A cultural analysis of the convergence of values across luxury and sustainability provides a pathway towards this outcome.

Table 4.1 Common values intrinsic to both luxury and sustainability

Common values	Luxury perspective	Sustainability perspective
High quality	The best marketplace proposition	Better quality limits make-take-waste cycle of consumption. Decreases overall product lifecycle waste
Excellence	Maintenance of brand equity	The satisfaction of consumption needs and considerate use of natural resources
Durability	Lasting brand experience	Indefinite existence of the planet
Timelessness	Dream status of luxury	The value of environment beyond material consumption
Heritage and long history	The long-held mastery traditions passed down the generations	Keeping natural environment for future generations
Rarity	Rare expensive resources	Protection of non-replenishable natural resources
Beauty	The finest art forms embedded in handcraft goods and exclusive experiences	The matchless beauty of the nature

4.5 The Case for Cultural Analysis of Luxury and Sustainability

Geertz provided grounding for socio-cultural research by arguing that "culture is not a power, something to which social events, behaviors, institutions, or processes can be causally attributed; it is a context, something within which they can be intelligibly—that is, thickly—described" (1983, p. 14). Cultural analysis uncovers dimensions of ideological representations, such as norms, rituals, language, symbols, and artefacts, that embed the ideology valued by a common group of people (Fischer, 2007; McCracken, 1986). Values are foundational to these dimensions and provide an opportunity to investigate links between the seemingly disparate concepts of luxury and sustainability. Values that are not necessarily identifiable but can be made evident through analysis of symbols, language, and norms (Fischer, 2007).

Both fashion and luxury discourses are socially constructed concepts that offer multiple dominant cultural meanings for consumers to interpret at an individual level (Roper et al., 2013; Thompson & Haytko, 1997). In emerging markets, for instance, fashion helps to express three aspects of consumer identity: an instrumental approach to self-construction, the balance between individualization and conformism, and the contradiction between globalized aspirations versus local limitations with past background (Kravets & Sandikci, 2014). Another example of cultural analysis is found in how the iconic status of shapewear—clothing that compresses and enhances the human body—is explained by its ability to maintain the material association and symbolism of both female oppression and empowerment throughout five centuries (Zanette & Scaraboto, 2019). Additionally, Davis' (1989) work explains how clothing can subtly demonstrate social identity and encode the status ambivalence of democracy and distinction through the use of two examples: the little black dress juxtaposing "mistresses–maids" associations, and jeans' working-class origin opposing the consumer-oriented middle class.

The same approach of multi-layered cultural meanings applies to sustainability. The inclusion of culture into sustainable development helps to overcome the limitations of individual particularism and acknowledges intergroup dynamics as well as historically evolving meanings of concepts (Proctor, 1998). Similar to luxury, sustainability is socially constructed and relies on the use of specific language, norms, symbols, and practices (Soini & Birkeland, 2014). An in-depth study of green consumers exposes common environmental values, shared practices of recycling and organic food consumption, and struggles of social relationships with those who do not share the same moral beliefs (Connolly & Prothero, 2008). In the case of solar heating systems, the placement of heaters in Swedish households depends on the combination of three culturally specific aspects: the perception of house and home, male or female spaces, and public or private spaces (Henning, 2004). Discourse analysis of academic literature on cultural sustainability reveals the range of ideologies from conservatism to liberalism to communitarianism and environmentalism and many value systems it incorporates (Soini & Birkeland, 2014).

These examples demonstrate multiple ways in which sustainability can be understood.

Sustainable consumption also includes symbolic and culturally situated meanings. Consumers used to hold a "darker, colder" image of sustainable future with less choice and comfort by associating it with "giving up and losing out" on certain consumption practices (Robins & Roberts, 1998: 30). Some consumers now see the sustainable discourse as a commodity to communicate their desired identity to others (Connolly & Prothero, 2003). For example, the Toyota Prius website uses the cultural appreciation of nature employing visual aesthetics, including birds, leaves, a girl with a dandelion, and a hand towards sun (de Burgh-Woodman & King, 2013). Consumers also perceive sustainability as an alternative path for satisfying self-interest, such as caring about their kids' future (Black & Cherrier, 2010). The mainstream culture moves away towards a greener perspective by "rationalizing" individual sustainable behaviour for both environmental and quality of life reasons (Prothero et al., 2010). A cultural analysis of sustainable luxury can reveal how firms link sustainability and luxury values.

4.6 Tiffany & Co.: A Cultural Analysis of Sustainable Luxury

While luxury is associated with stratification and class, and sustainability connotes social responsibility (Voyer & Beckham, 2014), identification of shared values provides opportunities for additional meanings at the convergence of luxury and sustainability. Sustainability researchers note the need to appeal not only consumers' rationality but also to emotional values in order to make sustainable options more attractive (Dolan, 2002; Martin & Väistö, 2016; Peattie & Peattie, 2009). Martin and Väistö (2016) argue that luxury electric vehicles offer consumers both environmentally conscious and hedonic values. The concept of luxury is not static; it opens up alternative meanings relevant for changing tastes of elites (Banister et al., 2020; Belk, 2020). A cultural analysis of these shared values can uncover the hedonic and sustainability benefits that appeal to sustainable luxury consumer. Tiffany & Co.'s values, as evidenced by its language, norms, and symbols, demonstrate how luxury firms can make a successful transition towards sustainability.

Tiffany is a legendary luxury brand that has recently embraced sustainability. Founded in 1837, Tiffany is among the oldest luxury companies and a globally recognized producer of artistic jewellery. Tiffany's brand equity is built on offering consumers a selection of finest diamonds and jewellery. A Tiffany Blue box that holds jewellery of a renowned design signifies the firm's historic luxury brand positioning. The colour of the distinctive Tiffany Blue packaging is under copyright (Tiffany & Co., 2021d). Even when empty, Tiffany presentation boxes signify luxury and retain value (EBay, 2021). The products are equally recognizable signifiers of luxury: an engagement ring with diamond mounted into a metal band from 1886, heart tag key ring from 1969 with the phrase "Please Return to Tiffany & Co.," key-shaped pendants from 2009, T collection of urban graphic style from 2014, all signify love and distinctive style associated with the Tiffany brand. Truman Capote's 1958 novel *Breakfast at Tiffany's* and the Audrey Hepburn film of the same name embedded the brand into mainstream twentieth-century culture.

Tiffany's sustainability claims are clearly evident on the company website. The webpage explicitly claims, "Sustainability is deeply ingrained in Tiffany's core values." Home page tabs labelled "Our Company" and "World of Tiffany" each include links to the "Sustainability" page, signifying the importance of linking the sustainability megatrend to the luxury brand. The "Sustainability" page includes brand's adherence to triple bottom line (Elkington, 1997): sustainable directives as well as sustainable goals, responsible sourcing, philanthropy, shared values, and wildlife protection. The "Future Focused" section states:

> At Tiffany, we celebrate love in our world and inspire love for our world. For over 25 years, we've been a leader in sustainable luxury, driving positive change across our three pillars of Product, People and Planet. Building on this legacy, we're proud to share our 2025 Sustainability Goals, which put a bold roadmap in place to guide us—and inspire our industry—towards a brighter future.

The "Sustainability" page includes links to sustainability reporting documents of the last fiscal year (2020) and previous reports starting with 2016. Company annual reports are often suspect as evidence of actual sustainability practices so corporations often rely on transparency to avoid accusations of greenwashing (Martin & Schouten, 2012). Including third-party validation of sustainability in corporate process and governance in annual reports adds rigour and validity to claims of sustainability (ibid.). Tiffany has been publishing its yearly sustainability reports since 2009 in accordance with the Global Reporting Initiative (GRI), and engaged KPMG, an independent registered public accounting firm for the Report of Independent Accountants (Tiffany & Co., 2020, 2021a). In 2016 *Forbes* published a quote from the brand's Chief Sustainability Officer:

> We'll continue to use the full power of the Tiffany brand to challenge the status quo, drive meaningful change and set the standard for sustainable luxury. Luxury and sustainability are, after all, deeply linked. Both are about heritage, quality and preserving beauty for generations to come. (McPherson, 2016)

Acknowledgement of shared values demonstrates an understanding of the importance of value congruence as Tiffany continues its gradual and consistent transition towards sustainability. These values are evidenced in the practices of the company's philanthropy and production.

The timeline of the company's sustainable efforts reveals activities linked to luxury/sustainable values of durability, timelessness, rarity, and beauty. Since 2002, The Tiffany & Co. Foundation has supported coral reef conservation and in 2018 committed $1 million to protect and restore Australian Great Barrier Reef, "one of the richest and most complex natural ecosystems on earth, and one of the most significant for biodiversity conservation" (Great Barrier Reef Foundation, 2018; UNESCO World Heritage Centre, 1981). In the 2000s Tiffany stopped buying gemstones from Myanmar, Burma (Tiffany & Co., 2021c), where the government is known for allowing human abuses in gemstone mines and relies on this trade to fund its activities (Human Rights Watch, 2008). The brand has extended its pioneering practices of responsible sourcing into diamond mining (Luxuo, 2020). In 2005 Tiffany became the first jeweller to join Earthworks' No Dirty Gold campaign which establishes social, human rights, and environmental standards for the extraction of gold. In 2011 Tiffany joined the United Nations Global Compact to align its practices with sustainability principles (Tiffany & Co., 2021c). From 2013 to 2019, Tiffany reduced their carbon emissions by 21%, and by 2020, 85% of the company's electricity use comes from clean sources, including renewable electricity and solar energy owned and generated by Tiffany (Tiffany & Co., 2021b). In 2020, 94% of the brand's Vietnamese workforce consisted of women hired from the local community.

The product range also exemplifies a shift towards sustainable values with the launch of the "Save the Wild" collection. All proceeds from the sales of animal shaped brooches and charms go to the Wildlife Conservation Network, which enabled the brand to donate $10 million to the organization. Even the iconic blue boxes and bags are now made with at least 50% recycled content (Ledru, 2020). In 2020 Tiffany became the first jewellery brand that achieved full transparency of its supply chain, tracing 100% of its rough diamonds to known mines and responsible suppliers (Kriser, 2020). In 2021 the brand made its first purchase of fair-mined certified artisanal gold (Tiffany & Co., 2021c). Tiffany's jewellery retains its material expression and the symbolic holder of (luxury) value (McCracken, 1986), while extending its reach towards sustainable ideology and inherent values.

Sustainable values of excellence, high quality, heritage, rarity, and beauty have become embedded in many other aspects of Tiffany's operations. The brand earned the sixth place on Barron's Annual Ranking of 100 Most Sustainable Companies in America, behind Best Buy, Agilent Technologies, Ecolab, Autodesk, and Voya Financial (Norton, 2021). The integration of language, norms, and symbols as markers of sustainable values has been intergraded into Tiffany's corporate culture. These practices demonstrate adoption of the environmental and social responsibility pillars of triple bottom line sustainability (Elkington, 1997). The third pillar, economic sustainability is evident the company's annual reports (*AnnualReports.com*, 2021).

The world's most powerful and valuable brands achieve their position by offering an innovative creative expression (Holt, 2012). Yet, in recent years Tiffany has been criticized for no longer making the cultural contribution that it was long known for (Danziger, 2020). The French luxury brand holding company Moët Hennessy Louis Vuitton (LVMH) acquired the brand in 2021 and is expected to provide financial support along with creativity and operational excellence, while Tiffany will in turn strengthen LVMH's presence in the American market (Danziger, 2019). LVMH's plans to revitalize the brand (Loeb, 2021); the next wave of Tiffany's cultural transformation will likely strengthen the luxury sustainability position.

New cultural expressions are built on existing brand assets, corresponding subcultures, and media myth (Holt, 2012). Tiffany successfully links its caring and loving image as manageable assets to cultural values of both luxury and sustainability. The advent of more responsible production and social responsibility links these values to the brand, while supporting corporate communication provide evidence of value congruence for the marketplace.

4.7 Conclusion and Future Research

We proposed that a cultural analysis as evidenced by language, symbols, and artefacts of a sustainable luxury brand can demonstrate congruent values between sustainability and luxury. The example of Tiffany & Co. shows how a luxury brand can embrace sustainability values while retaining luxury values. The look, fit, finish, and feel of luxury can be retained, and thus luxury values remain supported, even as the brand embraces sustainability. Yet, material and symbolic changes towards sustainability among luxury brands may not be obvious to consumers.

Research examining evidential material and symbolic changes in luxury brands moving towards sustainability would be useful to understand consumers' acceptance of value congruence. An investigation of luxury electric vehicle consumption is one option. Examining the luxury materializaion of other non-sustainable goods can provide important insights. For example, British jeweller Steven Webster's sterling silver straw called

Last Straw links his vanguard use of precious gemstones and love for parties with a durable cocktail straw (Frostrup, 2019). Daily, five million plastic straws are thrown into US garbage cans (Gibbens, 2019). This product, as a new cultural expression of luxury (Thomsen et al., 2020), suits luxury consumers who want to express their displeasure with plastic waste in upscale bar settings: the personally engraved silver straw offers functional durability, unlike paper, and luxury materiality, unlike stainless steel. It comes packed in a biodegradable pouch, and 10% of its sales proceeds go to Plastic Oceans International. "Last Straw" signals to others the pro-social identity of its user (Griskevicius et al., 2010). The name "Last Straw" solidifies the product's association with the values shared by sustainability and luxury: it is premium, caring, timeless, and beautiful. This product combines existing brand assets, corresponding subcultures and media myth to become associated with new cultural expressions (Holt, 2012).

Additional research into cultural analysis also lies in the area of universal sustainable symbols, symbols that can be attributed to the whole industry or broader area of life. One such symbol is the "Butterfly Mark" (described above), a certification awarded to sustainable luxury brands, akin to the Fair Trade label. The synergistic effect of all sustainable brands displaying the symbol strengthens the symbol's association with green practices and builds the Butterfly Mark's sign value, while also contributing to the development of a culturally situated symbol of sustainability (Mick, 1986; Soini & Birkeland, 2014). Further cultural analysis of the Butterfly Mark might investigate its potential in becoming a cultural symbol of our climate-aware times, similarly to the way the peace sign of the 1960s signalled opposition to nuclear weapons and the Vietnam War (Kolsbun & Sweeney, 2008). Research engaging a cultural analysis methodology can have impact on industry practitioners who look for culturally contextualized and actionable steps in both enhancing and communicating their sustainable efforts (Holt, 2020).

Consumer researchers are also urged to investigate how value congruence shapes the identity of sustainable luxury consumers. While the existence of educated elites and digital nomads are well described in the literature, research into the particular juxtaposition of sustainable consumers and luxury consumers is still limited (Atanasova & Eckhardt,

2021; Brooks, 2001; Currid-Halkett, 2017; Prothero et al., 2011). Rich and detailed delineation of consumers' life experiences offers insights for luxury firms looking to embrace the sustainability megatrend.

References

Aleksander, I. (2020). Sweatpants forever: How the fashion industry collapsed. *The New York Times*.
AnnualReports.com. (2021). Tiffany & Co. Accessed November 2, 2021, from https://www.annualreports.com/Company/tiffany-co
Atanasova, A., & Eckhardt, G. M. (2021). The broadening boundaries of materialism. *Marketing Theory*, 14705931211019077. https://doi.org/10.1177/14705931211019077
Athwal, N., Wells, V. K., Carrigan, M., et al. (2019). Sustainable luxury marketing: A synthesis and research agenda. *International Journal of Management Reviews, 21*(4), 405–426. https://doi.org/10.1111/ijmr.12195
Banister, E., Roper, S., & Potavanich, T. (2020). Consumers' practices of everyday luxury. *Journal of Business Research, 116*, 458–466. https://doi.org/10.1016/j.jbusres.2019.12.003
Belk, R. W. (2020). The changing notions of materialism and status in an increasingly dematerialized world. In: *Research handbook on luxury branding* (pp. 2–21).
Bellezza, S., Paharia, N., & Keinan, A. (2017). Conspicuous consumption of time: When busyness and lack of leisure time become a status symbol. *Journal of Consumer Research, 44*(1), 118–138. https://doi.org/10.1093/jcr/ucw076
Bendell, J., & Kleanthous, A. (2007). *Deeper luxury: Quality and style when the world matters*. World Wide Fund for Nature UK.
Black, I. R., & Cherrier, H. (2010). Anti-consumption as part of living a sustainable lifestyle: Daily practices, contextual motivations and subjective values. *Journal of Consumer Behaviour, 9*(6), 437–453. https://doi.org/10.1002/cb.337
Bowles, N. (2019, March 23). Human contact is now a luxury good. *The New York Times*.
Brooks, D. (2001). *Bobos in paradise: The new upper class and how they got there*. Touchstone.
Carson, R. (1962). *Silent Spring*. Houghton Mifflin Harcourt.

Cervellon, M.-C., & Drylie Carey, L. (2021). Luxury brands, consumer behaviour, and sustainability. In *Firms in the fashion industry* (pp. 87–95). Springer International Publishing. https://doi.org/10.1007/978-3-030-76255-1_6

Chatzidakis, A., & Lee, M. S. W. (2013). Anti-consumption as the study of reasons against. *Journal of Macromarketing, 33*(3), 190–203. https://doi.org/10.1177/0276146712462892

Cherrier, H., Black, I. R., & Lee, M. (2011). Intentional non-consumption for sustainability: Consumer resistance and/or anti-consumption? *European Journal of Marketing, 45*(11), 1757–1767. https://doi.org/10.1108/03090561111167397

Connolly, J., & Prothero, A. (2003). Sustainable consumption: Consumption, consumers and the commodity discourse. *Consumption, Markets and Culture, 6*(4), 275–291. https://doi.org/10.1080/1025386032000168311

Connolly, J., & Prothero, A. (2008). Green consumption: Life-politics, risk and contradictions. *Journal of Consumer Culture, 8*(1), 117–145. https://doi.org/10.1177/1469540507086422

Currid-Halkett, E. (2017). *The sum of small things: A theory of the aspirational class*. Princeton University Press. https://doi.org/10.1515/9781400884698.

Danziger, P. (2005). *Let them eat the cake*. Kaplan Trade.

Danziger, P. (2019). LVMH and Tiffany is a marriage made in luxury heaven. *Forbes*.

Danziger, P. (2020). If LVMH can't get out of Tiffany acquisition, as one lawyer believes, it may live to regret it. *Forbes*.

Davis, F. (1989). Of maids' uniforms and blue jeans: The drama of status ambivalences in clothing and fashion. *Qualitative Sociology, 12*(4), 337–355. https://doi.org/10.1007/BF00989396

de Burgh-Woodman, H., & King, D. (2013). Sustainability and the human/nature connection: A critical discourse analysis of being 'symbolically' sustainable. *Consumption, Markets and Culture, 16*(2), 145–168. https://doi.org/10.1080/10253866.2012.662834

Dean, A. (2018). *Everything is wrong: A search for order in the ethnometaphysical chaos of sustainable luxury fashion*.

DiDonato, T. E., & Jakubiak, B. K. (2016). Sustainable decisions signal sustainable relationships: How purchasing decisions affect perceptions and romantic attraction. *Journal of Social Psychology, 156*(1), 8–27. https://doi.org/10.1080/00224545.2015.1018858

Dolan, P. (2002). The sustainability of 'sustainable consumption'. *Journal of Macromarketing, 22*(2), 170–181. https://doi.org/10.1177/0276146702238220

4 Sustainable Luxury: A Framework for Meaning Through Value... 75

Dubois, B., & Duquesne, P. (1993). The market for luxury goods: Income versus culture. *European Journal of Marketing, 27*(1), 35–44.
EBay. (2021). *Tiffany & Co. jewelry boxes for sale*. Accessed November 1, 2021, from https://www.ebay.com/b/Tiffany-Co-Jewelry-Boxes/262017/bn_72172132
Elkington, J. (1997). *Cannibals with forks*. Capstone Publishing. https://doi.org/10.9774/gleaf.978-1-907643-44-6_24
Ernst & Young GL. (2017). *Why sustainable development goals should be in your business plan*. Accessed October 14, 2021, from https://www.ey.com/en_au/assurance/why-sustainable-development-goals-should-be-in-your-business-plan
Fifita, 'Ilaisaane, M. E., Seo, Y., Ko, E., et al. (2020). Fashioning organics: Wellbeing, sustainability, and status consumption practices. *Journal of Business Research, 117,* 664–671. https://doi.org/10.1016/j.jbusres.2019.01.005
Fischer, M. M. J. (2007). Culture and cultural analysis as experimental systems. *Cultural Anthropology, 22*(1), 1–65. https://doi.org/10.1525/can.2007.22.1.1
Fletcher, K. (2013). *Sustainable fashion and textiles: Design journeys*. Taylor and Francis. https://doi.org/10.4324/9781315857930
Frostrup, M. (2019). *True Tales of Luxury – Steven Webster*. Harrods.
Fuerst, F., & Shimizu, C. (2016). Green luxury goods? The economics of eco-labels in the Japanese housing market. *Journal of the Japanese and International Economies, 39,* 108–122. https://doi.org/10.1016/j.jjie.2016.01.003
Geertz, C. (1983). *Local knowledge: Further essays in interpretive anthropology*. Basic Books.
Geissdoerfer, M., Savaget, P., Bocken, N. M. P., et al. (2017). The circular economy – A new sustainability paradigm? *Journal of Cleaner Production, 143,* 757–768. https://doi.org/10.1016/j.jclepro.2016.12.048
Gibbens, S. (2019). *Plastic straw bans are spreading: Here's how they took over the world*. Accessed October 25, 2021, from https://www.nationalgeographic.com/environment/article/news-plastic-drinking-straw-history-ban
Goor, D., Ordabayeva, N., Keinan, A., et al. (2020). The impostor syndrome from luxury consumption. *Journal of Consumer Research, 46*(6), 1031–1051. https://doi.org/10.1093/jcr/ucz044
Great Barrier Reef Foundation. (2018). *The Tiffany & Co. Foundation Commits $1 Million to the Reef*.
Griskevicius, V., Tybur, J. M., & Van den Bergh, B. (2010). Going green to be seen: Status, reputation, and conspicuous conservation. *Journal of Personality and Social Psychology, 98*(3), 392–404. https://doi.org/10.1037/a0017346

Hansen, T., & Pollin, R. (2020). Economics and climate justice activism: Assessing the financial impact of the fossil fuel divestment movement. *Review of Social Economy*, 1–38. https://doi.org/10.1080/00346764.2020.1785539

Henning, A. (2004). Equal couples in equal houses: Cultural perspectives on Swedish solar and bio-pellet heating design. *Sustainable Architectures: Critical Explorations of Green Building Practice in Europe and North America, 9780203412*(2005), 89–103. https://doi.org/10.4324/9780203412800

Hoffman, A. J. (2018). The next phase of business sustainability. *SSRN Electronic Journal*. https://doi.org/10.2139/ssrn.3191035.

Holt, D. B. (2012). *Cultural brand strategy. Handbook of marketing strategy* (pp. 306–317). https://doi.org/10.4337/9781781005224.00029

Holt, D. (2020, September–October). Cultural innovation. *Harvard Business Review*, 1–11.

Human Rights Watch. (2008). *Burma's gem trade and human rights abuses*. Accessed November 2, 2021, from https://www.hrw.org/news/2008/07/29/burmas-gem-trade-and-human-rights-abuses#

IPCC U. (2021). *Climate change 2021: The physical science basis*. Cambridge University Press.

Jackson, T. (2005). Live better by consuming less? Is there a 'double dividend' in sustainable consumption? *Journal of Industrial Ecology, 9*(1–2), 19–36. https://doi.org/10.1162/1088198054084734

Joy, A., Sherry, J. F., Venkatesh, A., et al. (2012). Fast fashion, sustainability, and the ethical appeal of luxury brands. *Fashion Theory - Journal of Dress Body and Culture, 16*(3), 273–295. https://doi.org/10.2752/175174112X13340749707123

Kapferer, J.-N. (1997). Managing luxury brands. *Journal of Brand Management, 4*(4), 251–259. https://doi.org/10.1057/bm.1997.4

Kapferer, J.-N. (2010). All that glitters is not green: The challenge of sustainable luxury. *European Business Review (November–December)*, 40–45.

Kapferer, J.-N., & Michaut-Denizeau, A. (2014). Is luxury compatible with sustainability? Luxury consumers' viewpoint. *Journal of Brand Management, 21*(1), 1–22. https://doi.org/10.1057/bm.2013.19

Keinan, A., Crener, S., & Goor, D. (2020). Luxury and environmental responsibility. *Research Handbook on Luxury Branding, 300–322*. https://doi.org/10.4337/9781786436351.00031

Koch, M. (2021). GreenPrint survey finds consumers want to buy eco-friendly products, but don't know how to identify them. *Businesswire*, 50–51.

Kolsbun, K., & Sweeney, M. S. (2008). Peace: The biography of a symbol. *National Geographic*.

Kravets, O., & Sandikci, O. (2014). Competently ordinary: New middle class consumers in the emerging markets. *Journal of Marketing, 78*(4), 125–140. https://doi.org/10.1509/jm.12.0190

Kriser, H. (2020). Tiffany & Co. cements its leadership in diamond traceability. *Business Wire*, 18 August.

Kunz, J., May, S., & Schmidt, H. J. (2020). Sustainable luxury: Current status and perspectives for future research. *Business Research*. Springer International Publishing. https://doi.org/10.1007/s40685-020-00111-3.

Ledru, A. (2020). *A message from our CEO president and chief executive officer*.

Lee, M. S. W., & Ahn, C. S. Y. (2016). Anti-consumption, materialism, and consumer well-being. *Journal of Consumer Affairs, 50*(1), 18–47. https://doi.org/10.1111/joca.12089

Loeb, W. (2021). These changes could revamp The Tiffany & Co. *Forbes*.

Luxuo. (2020) How Tiffany & Co earned 4th most sustainable company on Barron's Annual List. Accessed October 27, 2021, from https://www.luxuo.com/homepage-slider/how-tiffany-co-earned-4th-most-sustainable-american-company-on-barrons-annual-list.html

Markard, J., Raven, R., & Truffer, B. (2012). Sustainability transitions: An emerging field of research and its prospects. *Research Policy, 41*(6), 955–967. https://doi.org/10.1016/j.respol.2012.02.013

Martin, D., & Schouten, J. W. (2012). *Sustainable marketing*. Pearson Prentice Hall.

Martin, D. M., & Väistö, T. (2016). Reducing the attitude-behavior gap in sustainable consumption: A theoretical proposition and the American electric vehicle market. *Review of Marketing Research, 13*, 193–213. https://doi.org/10.1108/S1548-643520160000013016.

McCracken, G. (1986). Culture and consumption: A theoretical account of the structure and movement of the cultural meaning of consumer goods. *Journal of Consumer Research, 13*(1), 71. https://doi.org/10.1086/209048.

McKibben, B. (2008). *The Bill McKibben reader: Pieces from an active life*. St. Martin's Griffin.

McPherson, S. (2016). Meet the woman driving sustainability and corporate responsibility at Tiffany & Co. *Forbes*.

Mick, D. G. (1986). Consumer research and semiotics: Exploring the morphology of signs, symbols, and significance. *Journal of Consumer Research, 13*(2), 196. https://doi.org/10.1086/209060

Mittelstaedt, J. D., Shultz, C. J., Kilbourne, W. E., et al. (2014). Sustainability as megatrend: Two schools of macromarketing thought. *Journal of Macromarketing, 34*(3), 253–264. https://doi.org/10.1177/0276146713520551

Mulkerrins, J. (2021). Frank Luntz: The man who came up with 'climate change'—and regrets it. *The Times*, 25 May.

Norgaard, K. M. (2011). Climate denial: Emotion, psychology, culture, and political economy. In J. S. Dryzek, R. B. Norgaard, & D. Schlosberg (Eds.), *Oxford handbook on climate change and society* (pp. 399–413). Oxford University Press.

Norton, L. P. (2021). The 100 most sustainable companies. *Barron's, 101*(7), 20–22, 24–26.

Ottman, J. A., Stafford, E. R., & Hartman, C. L. (2006). Avoiding green marketing myopia: Ways to improve consumer appeal for environmentally preferable products. *Environment: Science and Policy for Sustainable Development, 48*(5), 22–36. https://doi.org/10.3200/ENVT.48.5.22-36

Peattie, K., & Peattie, S. (2009). Social marketing: A pathway to consumption reduction? *Journal of Business Research, 62*(2), 260–268. https://doi.org/10.1016/j.jbusres.2008.01.033

Positive Luxury. (2022). *The Butterfly Mark*. Accessed January 17, 2022, from https://www.positiveluxury.com/butterfly-mark/

Proctor, J. D. (1998). The meaning of global environmental change: Retheorizing culture in human dimensions research. *Global Environmental Change, 8(3). Pergamon*, 227–248. https://doi.org/10.1016/S0959-3780(98)00006-5

Prothero, A., McDonagh, P., & Dobscha, S. (2010). Is green the new black? Reflections on a green commodity discourse. *Journal of Macromarketing, 30*(2), 147–159. https://doi.org/10.1177/0276146710361922

Prothero, A., Dobscha, S., Freund, J., et al. (2011). Sustainable consumption: Opportunities for consumer research and public policy. *Journal of Public Policy and Marketing, 30*(1), 31–38. https://doi.org/10.1509/jppm.30.1.31

Robins, N., & Roberts, S. (1998). Making sense of sustainable consumption. *Development, 41*(1), 28–36.

Roper, S., Caruana, R., Medway, D., et al. (2013). Constructing luxury brands: Exploring the role of consumer discourse. *European Journal of Marketing, 47*(3), 375–400. https://doi.org/10.1108/03090561311297382

Roux, D., & Izberk-Bilgin, E. (2018). Consumer resistance and power relationships in the marketplace. In E. J. Arnould & C. J. Thompson (Eds.), *Consumer culture theory* (pp. 295–317). Sage Publications.

Septianto, F., Seo, Y., & Errmann, A. C. (2021). Distinct effects of pride and gratitude appeals on sustainable luxury brands. *Journal of Business Ethics, 169*(2), 211–224. https://doi.org/10.1007/s10551-020-04484-7

Smith, N. C. (1990). *Morality and the market consumer pressure for corporate accountability*. Routledge. https://doi.org/10.4324/9781315743745

Soini, K., & Birkeland, I. (2014). Exploring the scientific discourse on cultural sustainability. *Geoforum, 51*, 213–223. https://doi.org/10.1016/J.GEOFORUM.2013.12.001

The Guardian. (2021). Harvard University will divest its $42bn endowment from all fossil fuels | Fossil fuel divestment. *The Guardian*.

The White House. (2021). Fact sheet: President Biden sets 2030 greenhouse gas pollution reduction target aimed at creating good-paying union jobs and securing U.S. Leadership on Clean Energy Technologies. *Whitehouse.Gov*: 6.

Thompson, C. J., & Haytko, D. L. (1997). Speaking of fashion: Consumers' uses of fashion discourses and the appropriation of countervailing cultural meanings. *Journal of Consumer Research, 24*(1), 15–42. https://doi.org/10.1086/209491

Thomsen, T. U., Holmqvist, J., von Wallpach, S., et al. (2020). Conceptualizing unconventional luxury. *Journal of Business Research, 116*, 441–445. https://doi.org/10.1016/j.jbusres.2020.01.058

Tiffany & Co. (2020). *Alignment to sustainability reporting frameworks contents*.

Tiffany & Co. (2021a). *Fiscal year 2020 sustainability performance and metrics*.

Tiffany & Co. (2021b). *Protecting our planet—Environmental philanthropy*. Accessed October 27, 2021, from https://www.tiffany.com.au/sustainability/the-planet/

Tiffany & Co. (2021c). *Sustainability timeline and milestones*. Accessed October 27, 2021, from https://www.tiffany.com.au/sustainability/timeline/

Tiffany & Co. (2021d). *Trademarks & Copyrights*. Accessed November 1, 2021, from https://www.tiffany.com.au/policy/trademarks-copyrights/

UK Government. (2008). *Climate Change Act 2008*: 1–108.

UNESCO World Heritage Centre. (1981). *Great Barrier Reef*. Accessed November 2, 2021, from http://whc.unesco.org/en/list/154

Voyer, B. G., & Beckham, D. (2014). Can sustainability be luxurious? A mixed-method investigation of implicit and explicit attitudes towards sustainable luxury consumption. *Advances in Consumer Research, 42*, 245–250.

Zanette, M. C., & Scaraboto, D. (2019). From the corset to Spanx: Shapewear as a marketplace icon. *Consumption, Markets and Culture, 22*(2), 183–199. https://doi.org/10.1080/10253866.2018.1497988

5

Sustainability, Saudi Arabia and Luxury Fashion Context: An Oxymoron or a New Way?

Sarah Ibrahim Alosaimi

5.1 Introduction

This chapter focuses on the Kingdom of Saudi Arabia (KSA) and its drive towards sustainability by further it focusing attention on the luxury fashion industry, which has seen a dramatic increase in uptake. Assomull (2020) has highlighted that Saudi consumers are treating luxury fashion increasingly like fast fashion, which implies that these garments and accessories are consumed often and in large quantities. Seeing as the KSA has recently undergone quite dramatic changes with the introduction of the Vision 2030, there is a need to explore not only what the government seeks to achieve with the introduction of their sustainability roadmap but also what implications might be for the luxury fashion industry, which is addressed in this chapter.

S. I. Alosaimi (✉)
Department of Fashion and Textile Design, Faculty of Art and Design, Princess Nourah Bint Abdul Rahman University, Riyadh, Saudi Arabia
e-mail: sialosaimi@pnu.edu.sa

© The Author(s), under exclusive license to Springer Nature Switzerland AG 2022
C. E. Henninger, N. K. Athwal (eds.), *Sustainable Luxury*, Palgrave Advances in Luxury, https://doi.org/10.1007/978-3-031-06928-4_5

Although the luxury fashion industry has seen increased interest, studies focusing on sustainable luxury remain limited (Athwal et al., 2019; Grazzini et al., 2021), with those that are available showing an obvious bias towards Western consumers. This could be seen as quite surprising, given the fact that the luxury market in the Middle East accounts for over 39% of the entire global luxury industry (Dogan-Sudas et al., 2019). The Saudi luxury market has reached $15.7 billion in 2020 and is forecasted to hit $23.6 by 2026, which is an increase of 7% between 2021 and 2026 (Expert Market Research, 2021). This highlights the importance of Middle Eastern luxury consumers (e.g., Dekhili et al., 2019; Hammad et al., 2019) as the Middle East has one of the world's most prosperous youth populations, where millennials are the next Arab luxury consumer crowd (Mishra et al., 2020).

Perhaps one explanation as to why there is only limited research into the Middle Eastern market could be accessibility and ability to conduct research, whilst a further might be the fact that Middle Eastern countries, which are guided by Islamic principle, are often seen as being anti-luxury. To explain, some authors (e.g., Alam et al., 2011; Abalkhail, 2021) indicate that luxury and Islam are not compatible as religion can have an impact on consumers' attitudes, whilst others contradict this opinion (Farrag & Hassan, 2015; Ashraf et al., 2017). What becomes apparent here is that the argument of whether or not luxury can be seen as accepted with Islamic principles might be country dependent. We see the complexity of sustainable luxury and Islamic principles emerge, which might become even more apparent when focusing on the KSA, which is not only Asia's fifth-largest country and the Arab world's second largest (Nurunnabi, 2017) but also the birthplace of Islam and home to the two holy cities and mosques, Mecca and Medina (important objects of Islamic pilgrimage) (Alsubaie et al., 2015; Nurunnabi, 2017; Abuljadail & Ha, 2019).

As alluded to, a paradox emerges in that consumption of luxury goods is a conservational issue in Islam, because Muslims are supposed to live thoughtful and modest lives, whereas luxury fashion consumption has long been related to extravagance (Ashraf et al., 2017). Yet, the KSA seems to be one of the largest luxury consumer markets, where the average luxury fashion consumer spends more than twice as much on luxury

5 Sustainability, Saudi Arabia and Luxury Fashion Context... 83

fashion as the average consumer in China (McKinsey & Company, 2020). Coming back to an earlier statement made, it seems as if Saudi consumers are treating luxury fashion in a similar manner as others do fast fashion (Assomull, 2020). This is especially prominent among millennial consumers, who form a majority of the Saudi population and have often been described as material collectors (Sadekar, 2020).

With increased luxury consumption practices come increased issues, including waste, recycling and pollution, which provides a key issue for the KSA. In 2016, the Saudi government introduced the Vision 2030 as a roadmap that guides the KSA to create a more diverse and sustainable economy, safeguards the environment by increasing the efficiency of waste management, establishes comprehensive recycling projects and reduces different types of pollution. In doing so, the country attempts to preserve the environment and natural resources, and fulfil Islamic, human and moral duties (Vision2030, 2016), which are now challenged due to overconsumption.

The interplay between having introduced the Vision 2030, which seeks a more sustainable future and increased luxury fashion consumption, makes the KSA an interesting context. To explain, the meaning of sustainability in this context needs to be further explored because there seems to be a paradox between the government's growing expectation on improving sustainable development and the luxury consumers' consumption practices. To encourage sustainable consumption in developing countries, it seems that there is a need to learn more about the practices and drivers to sustainable consumption within those countries (Hammad et al., 2019). Here, sustainable consumption refers to a resource-efficient and socially just method of consuming resources without jeopardising the needs of future generations (Hammad et al., 2019), which has become a widely discussed topic within consumer and marketing research (Henninger et al., 2016; Athwal et al., 2019). Therefore, it is critical to explore what sustainability and sustainable practices are within luxury fashion, which showcases the relevance of this chapter. Sustainability is not only a key priority for the Saudi government but is also critical in the luxury industry (Athwal et al., 2019; Donato et al., 2020), especially within the KSA, given the growth of the luxury market and the sales in the country, which are expected to continuously increase at around 7%

by 2026 despite the recent economic turmoil caused by COVID-19 (Expert Market Research, 2021; Fabre & Malauzat, 2021; IMARC Group, 2021). Particularly, there is an increased market demand for Western-branded luxury products in Gulf countries, including the KSA where the average spending on fashion reaches between $500 and $1600 per person (McKinsey & Company, 2020). The luxury industry is gaining public interest as a result of its high-profile background, which includes a wide range of environmental and social challenges that must be tackled, yet there seems to be a lack of consumer uptake for the reform (Kapferer, 2010; Cheah et al., 2016; Osburg et al., 2021). Unlike in developed countries where sustainable development and practices have been included in policies for a number of years, leading to consumers becoming more aware of sustainability issues and an increasingly positive attitude towards it (Dekhili et al., 2019; Kong et al., 2020); in the KSA, like any other developing country, the situation is different where the integration of sustainable development remains low, and thus authorities face major challenges in the adoption of sustainable development, including a lack of stakeholder interest and public awareness (Dekhili et al., 2019).

5.2 Background to Sustainability and Sustainability Practices in the KSA

5.2.1 Sustainability Through the Vision 2030

Sustainability is defined as meeting society's current needs without compromising on fulfilling those of future generations, by focusing on people, the planet and profit (Elkington, 1994). In the KSA, the Vision 2030 was introduced in 2016 and acts as a roadmap that guides the country to create a more diverse and sustainable economy, safeguard the environment by increasing the efficiency of waste management, establishing comprehensive recycling projects and reducing different types of pollution. In doing so, the country attempts to preserve environmental and natural resources; fulfil Islamic, human and moral duties; and secure future and current generations quality of everyday lives (Vision2030, 2016; Surf & Mostafa, 2017).

Thus, a key focus is on implementing a long-term strategy, which focuses on sustainable growth and development, thereby fostering diversity and moving the country towards becoming the centre of foreign trade (Alshuwaikhat & Mohammed, 2017; Rana & Suliman, 2018). The KSA seeks to reduce its dependency on oil and develops more environmentally friendly ways to conduct business (Rana & Suliman, 2018). As a result, the country has invested in a variety of sectors, including the (luxury) fashion industry, thereby fostering entrepreneurship and the creation of new market opportunities. Yet, a challenge is a fact that the (luxury) fashion industry is not only resource intensive but also one of the most polluting ones globally (Brown, 2020; Clark, 2020). How the luxury fashion industry can promote a sustainable outlook and support the Vision 2030 is addressed later in this chapter.

The Saudi government has adopted different efforts to promote sustainability through the Vision 2030, where different ministries, organisations and government bodies have begun restructuring to reconcile their practices and roles with the needs of this period (Alshuwaikhat & Mohammed, 2017; Rana & Suliman, 2018). The first step is to establish governance to oversee the plan's execution, and the second focuses on improving the country's strategies and incorporating them into long-term sustainability targets (Alshuwaikhat & Mohammed, 2017; Rana & Suliman, 2018). There are many promises in both the Vision 2030 and the government to reposition the KSA on the road to sustainability and guarantee that the KSA is included within the top 20 environmentally friendly nations globally and that the KSA's sustainable development targets are met by 2030 (Blaisi, 2019). Examples of these commitments include the Green Riyadh and the Green KSA projects, which seek to provide more greenery (e.g., trees and flowers) to cover the country and thus improve the overall air quality, lower the carbon dioxide emissions and improve impurity levels (Haddad, 2020; Saudi Press Agency, 2021a).

Additionally, the Saudi government has made a significant investment in different projects, including constructing and improving waste management and recycling facilities (Blaisi, 2019); providing high-quality sustainable goods and services; achieving water protection (Open Access

Government, 2021); building new urban societies and cities without noise or environmental pollution; creating more pedestrianised areas (Saudi Press Agency, 2021b); developing energy stream diversification, including renewable and alternative energy; decreasing the emissions of buried gases from landfills and turning them into energy; and protecting wildlife and plants by restricting hunting for wild birds and animals, as well as trade in endangered species (Unified National Platform, 2021). A majority of these suggestions are welcomed by the population and taken up, as there is a positive impact that becomes immediately obvious. Yet, when it comes to luxury fashion and its consumption, we see resistance, which is explained in the next section.

5.2.2 Sustainability and Luxury Fashion

As indicated, in an attempt to move away from its oil dependency, the KSA has heavily invested in other sectors, including the luxury fashion industry, by creating the Fashion Commission, which was established in 2020. The aim of the Fashion Commission is to nurture the growing industry, whilst 'it also seeks to enable the development of a fashion industry that is sustainable, inclusive, fully integrated along the value chain, and maximizes local talent and experience' (Arab News, 2021). Thus, it is taking an active role in promoting and fostering sustainable and ethical fashion innovations that can enable the KSA to become a global leader. It is commissioned to evaluate and report on the effect of various sustainability measures may have on local brands and entrepreneurs, as well as disseminate examples of best practices and share information on how to create more sustainable business models (Arab News, 2021; Fashion Commission, 2021; Mille, 2021). Examples of these sustainable business models include Sadeem, which focuses on cradle-to-cradle design (Assomull, 2019), or Abadia, which creates timeless design pieces and fosters local artisanry (Westernoff, 2019).

Saudi born 2020 Fashion Forward winner Yousef Akbar is a further forerunner for sustainable fashion, highlighting that whilst the country is changing through the Vision 2030 and people are becoming more environmentally conscious, the designer feels there is still a lot that needs to be done in order to make the industry as a whole more sustainable. In an interview with Saja Elmishri, Akbar insists that the two things that need to be further improved when it comes to sustainability are the education of both consumers and designers alike to create and buy into this type of fashion, as well as government support to facilitate this type of education (Curated Today, 2021). The notion of heritage, craftsmanship and artisanry is further reflected in the way the Fashion Commission has advertised sustainability to the public. Their aim is to foster and upgrade the fashion sector in the KSA through culture, promoting national heritage and identity in addition to achieving a positive impact on the national economy and thus creating a more sustainable economy (Fashion Commission, 2021).

It is noteworthy that what is seen as sustainable might differ, depending on the brand, designer and/or consumers' perceptions. The luxury industry has in the past been criticised, as often questionable materials (e.g., animal skins) or precious materials (e.g., gold) are used within the production process, which raises ethical concerns in relation to animal rights, but also in terms of exploitation of natural resources and the treatment of humans in the process of gaining these raw materials (Athwal et al., 2019; Luo et al. 2021). Moreover, a majority of the fashion production processes are based in low labour cost countries, which are often not geographically close to the country in which the products are consumed. This further provides environmental constraints, as goods that are imported and/or raw materials created come at an environmental (e.g., CO_2 emissions) and social cost (e.g., lack of living wage and unsafe working conditions) (Henninger et al., 2016; Brydges & Hanlon, 2020; Mukendi et al., 2020). Within the KSA currently, the fashion manufacturing industry is at a crossroads: 'there is a need for more employees as opportunities expand, but it is a youthful nation where industrial jobs in a lower-paying sector than oil are not always wanted' (Textile Journal, 2018). Although the country has a growing textile and fashion sector, a majority of the raw materials and/or garments are still imported (ibid).

Sustainability within the fashion industry was not explicitly advertised to the public through the KSA vision and the Fashion Commission, which mainly focuses on promoting national heritage and identity in addition to achieving impact on the national economy (Fashion

Commission, 2021). Therefore, sustainability remains a recent phenomenon in the KSA that has yet to be thoroughly explored, where Saudi consumers may not be aware of the Commission's mission of developing sustainable fashion or even associate it with their luxury fashion consumption. Therefore, it is critical to explore what sustainability and sustainable practices mean within luxury fashion, showcasing the relevance of this chapter.

5.2.3 Sustainability Practices in the KSA

Diverse perspectives exist on what sustainability entails and how it might be accomplished (Alshuwaikhat & Mohammed, 2017). Particularly, a careful review of existing literature on said field revealed that within existing Saudi consumer behaviour literature some contradictions are recognised. Some studies highlighted that sustainability is not practised or even exists within the KSA consumers (e.g., Assad, 2006, 2008; Dekhili et al., 2019). For example, there is a growing concern in the KSA about the negative effects of excessive consumption (Assad, 2006, 2008; Mohamad & Asfour, 2020; Sadekar, 2020). To illustrate, second-hand consumption in the KSA is not seen as a feasible option, as used fashion items are for charity and not for sale or not to be worn by the majority of Saudis (Al-fawaz, 2014; Assomull, 2020). Saudi consumers pack their old clothes and transport them to charitable societies such as the International Islamic Relief Organisation, where they are distributed to the poor all over the world (Al-fawaz, 2014; Assomull, 2020).

Additionally, Dekhili et al. (2019), for example, emphasised that sustainability information negatively impacts Saudi consumers' liking of luxury fashion items and lower the perceived quality. Luxury and social and environmental issues were viewed as incompatible by Saudi consumers, and thus, communicating that luxury products are made of alternative materials (e.g., vegetable leather) leads them to experience mental inconsistency (Festinger, 1962), evaluate the product quality negatively and reject it (Dekhili et al., 2019). Dekhili et al. (2019) provided a religious explanation as Muslims believe that nature was created by God alone and that only God can affect it. The Quran has around 500 verses that provide Muslims with environmental counsel (Achabou et al., 2021). God is believed to predetermine each person's future, rendering humans unaccountable for their actions' repercussions (Tsalikis & Lassar, 2009; Achabou et al., 2021); therefore, Saudi consumers may downplay their role in environmental conservation. This lack of consumer interest and awareness, in turn, confirms the major challenges that the country is facing in the adoption of sustainable development. Questions also aroused here are: how could this perception be changed? and what would the government and brands need to do to make sustainable consumption acceptable in the future?

On the other hand, other literature discusses some consumer practices that can be linked to sustainability. Whilst literature on Western consumers discusses consumers' rising ecological consciousness and underlines the significant gap between what customers profess to think and their actual behaviour (e.g., Joergens, 2006; Niinimäki, 2010), within the Saudi context the literature suggested different paradoxical behaviours exist between consumers' actual practices and their awareness of sustainable consumption. Saudi males, for example, are obliged to dress conservatively by wearing the traditional Saudi dress of loose white robes, the national 'thobe,' the headcover 'ghutra' and the black head-band 'iqal' in the workplace, schools and for different occasions and on a daily basis, regardless of their occupation, age or standing (Buchele, 2008; Gorney, 2016; Abdulaziz, 2019). Moreover, from an Islamic perspective, males are prohibited from wearing some luxury items (e.g., silk and jewellery) (Al-Mutawa, 2016; Abdulaziz, 2019), which means they have limited fashion consumption that is shaped by strong Islamic norms that require Saudis to dress modestly. Although this type of consumer practice seems to be driven by their belief system that reminds them to adhere to the moderate Islamic dress code, it can be considered as a sustainable practice, where they may not need multiple of those outfits as they look similar and do not change with the continuing fashion trend shifts.

While in developed countries globally luxury consumers have increasingly raised their expectations for environmental and social sustainability (De Angelis et al., 2020), there is a dearth of understanding among Saudi customers about sustainable clothing and its significance in environmental protection (Algahni & Al-Dabbagh, 2020). A study conducted on

Saudi mothers' awareness of sustainable fabrics by Berry et al. (2020) found that they purchase eco-friendly clothing for their children and inspect the eco-friendly labels not because they are environmentally friendly but because they look for clothing that protects their children against the harms of unsafe fabrics. While some brands communicate their sustainability information and efforts, this may have negative interpretations within Saudi consumers such as lack of quality perception of luxury items that are made of alternative materials or it might be back translated into the fact that eco-friendly clothing is safer for children. Both Berry et al.'s (2020) and Dekhili et al.'s (2019) findings highlighted the need to promote the notion of sustainability among Saudi consumers because those studies found Saudis to be less educated on what sustainability entails. This emphasised the government's need to take a step further and educate consumers to be able to comprehend and accept what brands communicate about their sustainability efforts and accept sustainable consumption practices.

5.3 Conclusion and Implications

Sustainability has emerged as a significant concern within the KSA because of the interplay between the Vision 2030, which seeks a more sustainable future and the increased luxury fashion consumption, where the country faces challenges in its mission. This chapter has explored sustainability issues within the Vision 2030 and its implementation, from the luxury fashion industry perspective. It also highlights the challenges the country may face in its mission to achieve the Vision 2030, where there is an increase in spending and subsequent luxury consumption, as well as sustainably questionable raw materials associated with the luxury fashion industry.

Although the country strives to establish environmentally friendly economic practices and invest in the fashion industry to move towards sustainability, still some sustainability challenges remain because sustainable development depends on how issues surrounding it are interpreted. Whilst there are some efforts to nurture the growth of the fashion industry and move away from the oil industry (see Fig. 5.1), the fashion

5 Sustainability, Saudi Arabia and Luxury Fashion Context... 91

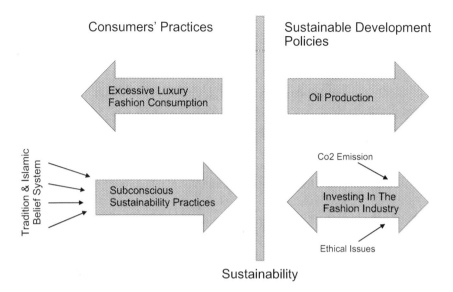

Fig. 5.1 Paradoxical framework for sustainability and luxury within the Saudi context

industry is not that much better since it is renowned as the second-greatest polluter behind the oil sector because of its significant carbon emissions, massive volumes of landfill debris, wastewater and low-quality working conditions (Colucci et al., 2020). The situation is quite similar to metaphorically speak 'sitting in a fence.' On the one hand, the country has started to move forward and can be more sustainable especially if sustainability is interpreted as not relying on oil. On the other hand, a question regarding sustainable development in the country remains because there are still hidden issues of sustainability that are not visible associated with the luxury fashion industry. Rather than relying on oil, some issues must be considered including the use of questionable materials such as leather or silk, which are animal products and have ethical issues, some fashion products use oil in their reproduction such as polyester, and CO_2 emission whether items are produced in the KSA or imported. Therefore, while some sustainability issues are sorted and some steps are taken towards sustainability, there are other issues that remain because the issue is not about the homegrown luxury fashion industry but rather consider different areas of hidden sustainability aspects that need to be incorporated.

Additionally, there is a gap between the government policies and efforts and consumers perception and practices of sustainability. Therefore, a paradox emerges between the government's growing expectation on improving sustainable development and luxury consumers' consumption practices, as illustrated in Fig. 5.1. Even when consumers consider being sustainable, it is difficult because they might have limited knowledge of what sustainability entails in the first place. Excessive consumption exists besides unconscious sustainable consumption practices. There are some limited types of sustainability practices that have been recognised among Saudi fashion consumers despite the lack of consumer awareness, yet it is still not clear what sustainability means for them. It is also obvious that some existing sustainable practices are driven by either Saudi tradition (e.g., traditional Saudi male dress) or Islamic belief system and norms (e.g., used fashion items given to charity, males not consuming jewellery and dress modestly). Sustainability is subconsciously practised through religion by following the rules prescribed in the Islamic belief system and trying to be a good Muslim in following God's instruction—which implies a key implication discussed in the following paragraph. Therefore, sustainable development and sustainable practices are overwhelmingly seen.

Therefore, this chapter suggests reconsidering policy improvements towards the Vision 2030 in the KSA, allowing both luxury fashion brands and consumers to follow specific guidelines that would assist them in behaving and consuming sustainably. Since the country strives for sustainability, it is suggested that the government introduce a set of rules and regulations concerning communication strategies to reach a wider audience. These rules can clearly outline criteria and implementation for sustainable practices not only to local luxury fashion brands but also to consumers. The government needs to educate both consumers and local fashion brands, regarding what sustainability means and how to create and consume environmentally friendly fashion items. As consumers are unconsciously behaving sustainably due to religious beliefs and norms, it is suggested to the government and brands to link their communication strategies and education programme to religion, which could facilitate the process of education and adoption of sustainability practices. Both the KSA government and luxury fashion brands should also not overlook the role of religion in major Muslim nations such as the KSA, where Islam is a strong factor that influences consumer decisions.

Besides educating Saudi local fashion brands about environmental challenges related to their work, brands also require assistance in developing strategies to reduce their brands' carbon emissions and move them towards a sustainable path. Those brands in turn need to hold educational seminars and workshops for their markets and managers to bring a great level of awareness that can be clearly communicated to consumers. Key managerial implications also for luxury brands marketers is that in order to sell luxury fashion items that are known as sustainable, it is recommended to apply an acculturation approach to Saudi society by undergoing a cultural assimilation process to better understand how their communication of sustainable fashion is back translated within Saudi consumers. Understanding the various viewpoints and unconscious practices that exist is vital for marketers, who can understand what makes luxury fashion sustainable from the consumers perspective and try to package it in a way that consumers understand better and promote it through different channels, which eventually helps to enhance sustainable development.

As this review reveals that there is a lack of awareness and limits of sustainability practices, future work may consider exploring how best practices developed outside of the KSA can be adapted to better suit local environmental and cultural conditions. Research can also explore the effectiveness of implementing specific sustainable fashion consumption practices (e.g., swapping, renting and second-hand buying). Future research should aim to identify what sustainability means for consumers to help the government to develop their education strategies to make sustainable consumption acceptable in the future.

References

Abalkhail, T. S. (2021). The impact of religiosity on luxury brand consumption: The case of Saudi consumers. *Journal of Islamic Marketing, 12*(4), 763–775.

Abdulaziz, D. (2019). Saudi women are breaking free from the black abaya. *The Wall Street Journal*. [Online] Accessed June 10, 2021, from https://www.wsj.com/articles/saudi-women-are-breaking-free-from-the-black-abaya-11570008601

Abuljadail, M. H., & Ha, L. (2019). What do marketers post on brands' Facebook pages in Islamic countries? An exploratory study of local and global brands in Saudi Arabia. *Journal of Islamic Marketing, 10*(4), 1272–1287.

Achabou, M. A., Dekhili, S., & Hamdoun, M. (2021). How the country of origin cue affects consumer preference in the case of ecological products: An empirical study in two developing countries. *Journal of Strategic Marketing*, 1–17.

Alam, S. S., Rohani, M., & Badrul, H. (2011). Is religiosity an important determinant on Muslim consumer behaviour in Malaysia? *Journal of Islamic Marketing, 2*(1), 83–96.

Al-fawaz, N. (2014). Secondhand clothes not an option for Saudis. *Arab News*. [Online] Accessed June 9, 2021, from https://www.arabnews.com/saudi-arabia/news/660336

Algahni, H., & Al-Dabbagh, M. A. (2020). Assessing the impact of social media in the consumer trend towards sustainable clothing. *Fibres and Textiles, 3*.

Al-Mutawa, F. S. (2016). Negotiating Muslim masculinity: Androgynous spaces within feminized fashion. *Journal of Fashion Marketing and Management, 20*(1), 19–33.

Alshuwaikhat, H. M., & Mohammed, I. (2017). Sustainability matters in national development visions-evidence from Saudi Arabia's vision for 2030. *Sustainability*, 1–15.

Alsubaie, H., Valenzuela, F. R., & Adapa, S. (2015). The advent of western-style shopping centres and changes in Saudi women's purchasing behaviour. In H. El-Gohary & R. Eid (Eds.), *Emerging research on Islamic marketing and tourism in the global economy* (pp. 19–41). Business Science Reference.

Arab News. (2021). Saudi Arabia's fashion Commission finalizes sector development strategy. *Arab News*. [Online] Accessed May 6, 2021, from https://www.arabnews.com/node/1852566/amp

Ashraf, S., Hafeez, M. H., Yaseen, A., & Naqvi, A. (2017). Do they care what they believe? Exploring the impact of religiosity on intention to purchase luxury products. *Pakistan Journal of Commerce and Social Sciences, 11*(2), 428–447.

Assad, S. W. (2006). Facing the challenges of consumerism in Saudi Arabia. *Journal of King Saud University, 19*(1), 1–20.

Assad, S. W. (2008). The rise of consumerism in Saudi Arabian society. *International Journal of Commerce and Management, 17*(1/2), 73–104.

Assomull, S. (2019). How Arab fashion is waking up to sustainability. *Arab News* (online) Accessed June 2, 2021, from https://www.arabnews.com/node/1588041/lifestyle

Assomull, S. (2020). Long snubbed, secondhand luxury catches on in the Middle East. *Vogue Business*. [Online] Accessed October 30, 2020, from https://www.voguebusiness.com/sustainability/long-snubbed-secondhand-luxury-catches-on-in-the-middle-east

Athwal, N., Wells, V. K., Carrigan, M., & Henninger, C. E. (2019). Sustainable luxury marketing: A synthesis and research agenda. *International Journal of Management Reviews, 21*(4), 405–426.

Berry, H. S., Ismail, R. K., Al-Daadi, S. E., Badr, S. I. O., Mesbah, Y. O., & Dabbagh, M. A. (2020). Measuring Saudi mothers' awareness of sustainable children's clothing. *Open Journal of Social Sciences, 8*(11), 244–262.

Blaisi, N. I. (2019). Construction and demolition waste management in Saudi Arabia: Current practice and roadmap for sustainable management. *Journal of Cleaner Production, 221*, 167–175.

Brown, H. (2020). Coronavirus and sustainability: How will fashion respond? *Drapers*. https://www.drapersonline.com/business-operations/coronavirus-and-sustainability-how-will-fashion-respond/7040147.article, 27/06/2020.

Brydges, T., & Hanlon, M. (2020). Garment worker rights and the fashion industry's response to COVID-19. *Dialogues in Human Geography, 10*(2), 195–198.

Buchele, N. (2008). *Saudi Arabia-culture smart! The essential guide to customs & culture*. Kuperard.

Cheah, I., Zainol, Z., & Phau, I. (2016). Conceptualizing country-of-ingredient authenticity of luxury brands. *Journal of Business Research, 96*(12), 5819–5826.

Clark, R. (2020). Global fashion exchange launches new digital swapping systems. *Forbes*. https://www.forbes.com/sites/roddyclarke/2020/05/27/global-fashion-exchange-launches-new-digital-swapping-platform/, 27/06/2020.

Colucci, M., Tuan, A., & Visentin, M. (2020). An empirical investigation of the drivers of CSR talk and walk in the fashion industry. *Journal of Cleaner Production, 248*, 119200.

Curated Today. (2021) Ethical fashion culture on the rise in the KSA feat. Yousef Akbar. *Curated Today* (online). https://curatedtoday.com/ethical-fashion-culture-on-the-rise-in-ksa/, 07/06/2021.

De Angelis, M., Amatulli, C., & Pinato, G. (2020). Sustainability in the apparel industry: The role of consumers' fashion consciousness. In S. S. Muthu & M. A. Gardetti (Eds.), *Sustainability in the textile and apparel industries* (pp. 19–33). Springer.

Dekhili, S., Achabou, M. A., & Alharbi, F. (2019). Could sustainability improve the promotion of luxury products? *European Business Review, 31*(4), 488–511.

Dogan-Sudas, H., Kara, A., & Cabuk, S. (2019). The role of sustainable environment attributes in luxury product perceptions: Evidence from an emerging market. *Journal of Transnational Management, 24*(1), 3–20.

Donato, C., De Angelis, M., & Amatulli, C. (2020). Sustainable luxury: The effect of corporate social responsibility strategy on luxury consumption motivations. In I. Cantista & T. Sádaba (Eds.), *Understanding luxury fashion* (pp. 123–146). Palgrave Macmillan.

Elkington, J. (1994). Towards the sustainability corporation: Win-win-win business strategies for sustainable development. *California Management Review*, Winter, 90–100.

Expert Market Research. (2021). Saudi Arabia luxury market outlook. *Expert Market Research*. [Online] Accessed May 19, 2021, from https://www.expertmarketresearch.com/reports/saudi-arabia-luxury-market

Fabre, C., & Malauzat, A. (2021). Assessing the impact of 2020 on the GCC luxury goods market. *Bain & Company*. [Online] Accessed May 19, 2021, from https://www.bain.com/insights/assessing-the-impact-of-2020-on-the-gcc-goods-luxury-market/

Farrag, D. A., & Hassan, M. (2015). The influence of religiosity on Egyptian Muslim youths' attitude towards fashion. *Journal of Islamic Marketing, 6*(1), 95–108.

Fashion Commission. (2021). About fashion | fashion Commission. *Fashion Commission*. [Online] Accessed May 6, 2021, from https://fashion.moc.gov.sa/en/about-fashion

Festinger, L. (1962). *A theory of cognitive dissonance*. Stanford University Press.

Gorney, C. (2016). The changing face of Saudi women. *National Geographic*. [Online] Accessed June 3, 2017, from http://ngm.nationalgeographic.com/2016/02/saudi-arabia-women-text

Grazzini, L., Acuti, D., & Aiello, G. (2021). Solving the puzzle of sustainable fashion consumption: The role of consumers' implicit attitudes and perceived warmth. *Journal of Cleaner Production, 287,* 125579.

Haddad, R. (2020). Saudi vision 2030 to plant 7.5 million trees in Riyadh. *About Her.* [Online] Accessed May 7, 2021, from https://www.abouther.com/node/31216/people/features/saudi-vision-2030-plant-75-million-trees-riyadh

Hammad, H., Muster, V., El-Bassiouny, N. M., & Schaefer, M. (2019). Status and sustainability: Can conspicuous motives foster sustainable consumption in newly industrialized countries? *Journal of Fashion Marketing and Management, 23*(4), 537–550.

Henninger, C. E., Alevizou, P. J., & Oates, C. J. (2016). What is sustainable fashion? *emerald.com, 20*(4), 400–416.

IMARC Group. (2021). Saudi Arabia luxury market: Industry trends, share, size, growth, opportunity and forecast 2021-2026. *IMARC Group.* [Online] Accessed May 19, 2021, from https://www.imarcgroup.com/saudi-arabia-luxury-market

Joergens, C. (2006). Ethical fashion: Myth or future trend? *Journal of Fashion Marketing and Management, 10*(3), 360–371.

Kapferer, J. N. (2010). All that glitters is not green: The challenge of sustainable luxury. *European Business Review, 2,* 40–54.

Kong, H. M., Witmaier, A., & Ko, E. (2020). Sustainability and social media communication: How consumers respond to marketing efforts of luxury and non-luxury fashion brands. *Journal of Business Research.*

Luo, S., Henninger, C. E., Le Normand, A., & Blazquez, M. (2021). Sustainable what…? The role of corporate websites in communicating material innovations in the luxury fashion industry. *Journal of Design, Business and Society, 7*(1), 83–103.

McKinsey & Company. (2020). *The state of fashion 2020.*

Mille. (2021) The future of Saudi Arabia's fashion scene is looking bright. *Mille.* [Online] Accessed May 7, 2021, from https://www.milleworld.com/burak-cakmak-saudi-arabia-fashion-commission/

Mishra, S., Jain, S., & Jham, V. (2020). Luxury rental purchase intention among millennials—A crossnational study. *Thunderbird International Business Review,* 1–14.

Mohamad, K. A., & Asfour, A. M. (2020). Is the business in Saudi market sustainable? Case study of Almarai. *IIUM Journal of Case Studies in Management, 11*(2), 7–14.

Mukendi, A., Davies, I., Glozer, S., & McDonagh, P. (2020). Sustainable fashion: Current and future research directions. *European Journal of Marketing, 54*(11), 2873–2909.

Niinimäki, K. (2010). Eco-clothing, consumer identity and ideology. *Sustainable Development, 18*(3), 150–162.

Nurunnabi, M. (2017). Transformation from an oil-based economy to a knowledge-based economy in Saudi Arabia: The direction of Saudi vision 2030. *Journal of the Knowledge Economy, 8*(2), 536–564.

Open Access Government. (2021). Advancing sustainability in the Kingdom of Saudi Arabia. *Open Access Government.* [Online] Accessed May 8, 2021, from https://www.openaccessgovernment.org/advancing-sustainability-in-the-kingdom-of-saudi-arabia/108355/

Osburg, V. S., Davies, I., Yoganathan, V., & McLeay, F. (2021). Perspectives, opportunities and tensions in ethical and sustainable luxury: Introduction to the thematic symposium. *Journal of Business Ethics, 169*(2), 201–210.

Rana, D., & Suliman, R. (2018). Green business: Sustainability within Saudi Vision 2030. *International Journal of Advance Study and Research Work, 1*(9), 4–14.

Sadekar, R. (2020). How are luxury goods perceived in the Gulf? *Hall & Partner.* [Online] Accessed June 2, 2020, from https://www.hallandpartners.com/luxury-perceptions-in-the-gulf

Saudi Press Agency. (2021a). Economist / his highness the Crown Prince announces the Saudi Green initiative and the Green Middle East initiative. *Saudi Press Agency.* [Online] Accessed May 10, 2021, from https://www.spa.gov.sa/2208368

Saudi Press Agency. (2021b). Economist/his highness the Crown Prince launches the 'The Line' project in Neom. *Saudi Press Agency.* [Online] Accessed May 10, 2021, from https://www.spa.gov.sa/viewstory.php?lang=ru&newsid=2177615

Surf, M. S., & Mostafa, L. A. (2017). Will the Saudi's 2030 vision raise the public awareness of sustainable practices? *Procedia Environmental Sciences, 37*, 514–527.

Textile Journal. (2018). Textile industry in Saudi Arabia. *Textile Journal* (online): https://kohantextiljournal.com/saudi-arabia-textile-market/, 24/06/2021.

Tsalikis, J., & Lassar, W. (2009). Measuring consumer perceptions of business ethical behaviour in two Muslim countries. *Journal of Business Ethics, 89*(1), 91–98.

Unified National Platform. (2021) Sustainable development. *Unified National Platform*. [Online] Accessed May 8, 2021, from https://www.my.gov.sa/wps/portal/snp/content/SDGPortal/!ut/p/z0/04_Sj9CPykssy0xPLMnMz0vMAfIjo8zi_QxdDTwMTQz93YMt3AwCzXyMg1wMAw0NLA31g1Pz9AuyHRUBEXub1w!!/

Vision2030. (2016). *Saudi vision 2030*. Kingdom of Saudi Arabia.

Westernoff, N. (2019). Meet the ethical Saudi brand that dresses Queen Rania. *Emirates Woman* (online). Accessed May 8, 2021, from https://emirateswoman.com/spotlight-on-sustainable-fashion-brand-abadia/

6

Towards Circular Luxury Entrepreneurship: A Saudi Female Entrepreneur Perspective

Rana Alblowi, Claudia E. Henninger, Rachel Parker-Strak, and Marta Blazquez

6.1 Introduction

Saudi female entrepreneurs have increased and seemingly taken initiative to create value in the Saudi luxury fashion market (Ramadan & Nsouli, 2021; Obaid, 2021b). This may be partially linked with the country's introduction of the Circular Economy strategy, which supports the fashion industry, with the focus on sustainability goals guided by the Vision 2030. Thus, Saudi Fashion Futures focus on sustainability, diversity, culture, innovation, and entrepreneurship (Arab News, 2021b).

To reiterate this further, Saudi Arabia has adopted a circular carbon economy strategy known as the circular carbon economy (Hamdan,

---

R. Alblowi (✉)
Department of Materials, The University of Manchester, Manchester, UK

Prince Sattam Bin Abdulaziz University, Al-Kharj, Saudi Arabia
e-mail: r.alblowi@psau.edu.sa

C. E. Henninger • R. Parker-Strak • M. Blazquez
Department of Materials, The University of Manchester, Manchester, UK

© The Author(s), under exclusive license to Springer Nature Switzerland AG 2022
C. E. Henninger, N. K. Athwal (eds.), *Sustainable Luxury*, Palgrave Advances in Luxury, https://doi.org/10.1007/978-3-031-06928-4_6

2021b; Malek, 2021; Kim et al., 2021b). The circular economy concept is based on a closed-loop economy (Brydges, 2021). Thus it is a model of production and consumption that emphasises the restoration and regeneration of ecosystems (Almulhim & Abubakar, 2021). It enables the utilisation of renewable energy for production and eliminates toxic chemicals while promoting the reuse and recycling of products (Niinimäki & Karell, 2020).

The Ellen MacArthur Foundation is often cited, as they have presented innovative ideas for business remodelling, product redesigning, and changing consumption patterns to ensure environmental sustainability (Ellen MacArthur Foundation, 2021). Thus far, the circular economy remains a rather recent phenomenon (Athwal et al., 2019) as governments and academia have only recently focused on the circular economy as a viable alternative for a sustainable future (Ki et al., 2020). Within Saudi Arabia, the Saudi Vision 2030 is supportive of this transformation and calls for practical implementation of circular economy principles.

Saudi Arabia has developed a conducive policy environment supported by numerous plans and programmes which support fashion industry entrepreneurs (Basahal, 2020) to build a circular economy. This chapter focuses on fashion entrepreneurs, more specifically female Saudi luxury fashion entrepreneurs, who have increased in numbers and are a driving force in creating value in the Saudi luxury fashion market (Kim et al., 2021b; Ramadan & Nsouli, 2021; Obaid, 2021b). Even though they are seen as a driving force, female entrepreneurs may not have an adequate understanding of what sustainability is and how to implement it within their "businesses" operations, seeing as "sustainability" is a relatively new phenomenon in Saudi Arabia. While the Ministry of Culture and the Fashion Commission have developed various strategies related to sustainability, there is a lack of proliferation of concrete steps and processes those female entrepreneurs can follow to be part of the circular economy. Within this chapter, sustainability and the circular economy are closely interlined and interconnected. Thus, sustainability is seen as the overarching goal that is fostered through circular economy strategies.

The fact that female Saudi luxury fashion entrepreneurs lack awareness may hinder them from fully capitalising on available resources and often not achieving their sustainability goals (Basahal, 2020). Few studies have investigated the effectiveness of public awareness, attitudes, and lifestyles in the transition to a circular economy (Almulhim & Abubakar, 2021). Moreover, there is a socio-cultural aspect to these challenges, as fashion entrepreneurs face a market that may not be ready to accept recycled or second-hand luxury goods, which form part of the sustainability strategy. The Saudi Vision 2030 and other governmental policies give female entrepreneurs direction and resources, but they do not consider the social stigma or bias that it may entail with recycled garments in the country. However, recycling materials is one aspect that is linked to having a circular economy strategy. According to Basahal (2020), there are almost no current studies focusing on the role of female entrepreneurs in supporting sustainability initiatives in the luxury fashion industry in Saudi Arabia. This is a gap in research and warrants an exploration of the current status of understanding and implementation of sustainability principles and circular economy strategies by Saudi female entrepreneurs.

This chapter, therefore, explores the challenges Saudi female entrepreneurs faced in successfully taking advantage of the Circular Economy Strategy from the Saudi Vision 2030. Thus, this chapter poses the following research aims:

1. To gain a better understanding of Saudi female fashion entrepreneurs' perceptions of the circular economy in light of the Saudi Vision 2030.
2. To evaluate the challenges Saudi female fashion entrepreneurs face in developing sustainable solutions in the luxury fashion market.

6.2 Literature Review

6.2.1 Sustainability in the Luxury Fashion Market

With the growth of globalisation and the increasing interlink and reliance of economies on each other, there has been a surge in luxury fashion consumption globally (Cabigiosu, 2020; Kim et al., 2021a), leading to an increased academic interest in developing markets. It is especially true in the Middle East, where people's high purchasing power capacity qualifies them as critical luxury fashion consumers (Centobelli et al., 2020b).

However, despite the market's rapid growth, the fashion industry is criticised for causing environmental damage and struggles to implement sustainability (Battle et al., 2018; Balconi et al., 2019; Henninger et al., 2021a). This industry is among the most polluting industries identified globally (Henninger et al., 2017; Balconi et al., 2019; Henninger et al., 2021b), which has been a cause of concern.

Within Saudi Arabia, this shift can also be observed, with the appearance of the new Arab audience—millennials—who are increasingly interested in sustainable options (Mishra et al., 2021). Millennials recognise luxury fashion companies as ones that have a worldwide reputation, excellent quality and innovation, and core competency (Ramadan & Nsouli, 2021), which has led to more research in the field of sustainable fashion technology and consumption in the middle east, especially in Saudi. Yet, whilst there is research on consumers, there is a gap addressing the role of entrepreneurship in generating innovative economic models and transforming the luxury fashion industry (Mishra et al., 2020; Almulhim & Abubakar, 2021), especially in the context of the Saudi Arabia (Ramadan & Mrad, 2017; Almulhim & Abubakar, 2021). Therefore, the fashion industry is on a path to transition as the emergence of a circular economy has made it possible to move away from linear processes like make-use-dispose to reuse, reutilise, and closed-loop approaches to consumption (Henninger et al., 2020).

6.2.2 Circular Economy and Luxury Fashion Industry

The circular economy is seen as a potential solution to enhancing sustainability in the fashion industry (Brydges, 2021), along with the investment in innovation that could enable the use of clean technology and processes in manufacturing. A circular economy presents a restorative economic model (Jain & Mishra, 2019), which can lead to overall environmental sustainability and community enhancement. By encouraging superior designing of products and utilisation of resources in an environmentally friendly manner, circular economy leads to better quality products at affordable costs (Ellen MacArthur Foundation, 2012). The benefits are multi-fold if the environmental regeneration and the prolonged lifecycle of products from a circular economy are considered (ibid.). While the concept of second-hand luxury fashion is aligned with sustainability goals (Henninger et al., 2016), it involves decoupling economic processes from using non-renewable resources and ensuring minimal wastage (Bocken et al., 2016). The circular economy emphasises the redesigning of products, re-engineering processes, and ensuring a focus on regeneration. One prominent focus of the circular economy is to ensure that products and materials are not wasted and instead kept in use for more extended periods, and as such, the second-hand garment industry could be complementary to the circular model, though not only resell but also swapping, renting, or other models (Hidalgo et al., 2019). The circular economy is seen as a potential solution to enhancing sustainability in the fashion industry (Centobelli et al., 2020a, b), along with the investment in innovation that could enable the use of clean technology and processes in manufacturing. There are various definitions of the circular economy in literature, from a systemic shift in the economy to a more straightforward combination of activities that focus on reducing, reusing, and recycling waste (Niinimäki, 2017). However, the circular economy is explicitly linked with economic prosperity brought about by sustainable development (Bertassini et al., 2021). A more expansive view of the circular economy also includes generating social equality and empowering the weaker sections of society (Geissdoerfer et al., 2017). However, academics have focused on circular economy in the last few decades, allowing for clarity on implementing the rather vague notion of sustainability (Murray et al., 2017).

The field of research in the circular economy is still emerging, with both the definition and concept of the circular economy expanding and becoming reconstructed with new evidence (Henninger et al., 2020). Several different theories around circular economy principles are based on sector or market focus (Henninger et al., 2021b). Therefore, it is critical to understand the circular economy's implications for the fashion industry, particularly the luxury fashion sector in Saudi Arabia, in this context. According to Niinimäki (2017), in a circular economy, the fashion industry is expected to focus on sustainability and reduce the long-term impact of its processes, through designing for longevity, service, material recovery, and reuse in manufacture. Niinimäki (2017) states that developing a sustainable business model for the luxury fashion industry requires focusing on radical innovations in design and technology and implementing new strategies that include all stakeholders, including the community and the environment. Moreover, the circular economy entails that the methods of using textiles need to transform at a fundamental level—that the fabrication, use, and disposal of materials are undertaken in new and innovative ways which are different from the traditional usage. Shirvanimoghaddam et al. (2020) postulate that the circular economy requires that supplies and finished products be retained for a prolonged period while maintaining value creation. They further elaborate that there is a need to think of innovative ways to repurpose the products once they have been used and utilise them in designing and developing new products. There are also specific aspects of the circular economy that can be employed in the fashion luxury market. For example, Henninger et al. (2021a) have found that companies can contribute to the circular economy by collecting their products back from their customers and repurposing them. The recollected material is used for innovative designing of new material, and hence it can be kept in use for a longer time, instead of being sent to landfills as waste (Ræbild & Bang, 2017). Another approach is developing synthetic raw materials and fibres that last longer, resist wear and tear, and thus increase longevity (Niinimäki & Karell, 2020). Research has also revealed that companies are investing in new technologies that can enable them to develop textiles and clothes that are more environmentally friendly and reduce the overall carbon footprint (Hvass & Pedersen, 2019). Though this may not be directly related to the circular economy concept, it adds to sustainability.

Similarly, the second-hand luxury fashion market can be considered to contribute to the circular economy. Second-hand clothes are bought by local vendors and sold back to merchants that can repurpose or redesign and resell these at a profit, or simply resell second-hand goods, often at a lower price (Moorhouse & Moorhouse, 2017). Similarly, researchers have explored the utility of creative marketing in campaigns that encourage clothes swapping between people and thus maintain that the clothes stay in use for a prolonged period (Henninger, 2021a). While there is considerable work done within the fashion industry to become environmentally friendly, there appears to be a lack of academic research that could explore the practical aspects of implementing a circular economy in the luxury fashion industry.

Moreover, an essential strategy that is inherently required for the circular economy to be effectively implemented is to create awareness around sustainability and the need for customers to be sensitive to the environmental impact of their consumption (Henninger et al., 2018). It is here that the role of the government becomes paramount, as it has the resources and the stake to ensure that both the entrepreneurs and the consumers are aware of sustainability and circular economy-related concepts. Therefore, the following section focuses on the Saudi government's role in generating sustainability in the fashion industry.

6.2.3 Role of Government in Saudi in Generating Sustainability in the Fashion Industry

Governments are critical in facilitating the ecosystem and establishing the regulatory framework within which businesses can operate (Centobelli et al., 2020a). In the context of Saudi Arabia, the government has taken a proactive role in supporting sustainability in businesses (Ramadan & Nsouli, 2021). It is evident in the provisions of the Saudi Vision 2030 that specifically pertains to the fashion industry's environmental and social sustainability and proposes a circular economy development (Hamdan, 2021a). The Government of Saudi Arabia and the Saudi Fashion Commission (under the Ministry of Culture) have committed to developing sustainability initiatives driven by the Saudi Vision 2030. Under its "100 Saudi Brands program" scheme, the Saudi Fashion Commission aims to empower and support local luxury brands to attain international standards (Arab News, 2021a). This programme provides support and skill development to designers and entrepreneurs in the fashion industry to take advantage of the circular economy, improve their sustainability, and develop leadership capabilities (Fashion Commission, 2021a). Additionally, the Ministry of Culture has a Fashion Futures Programme that focuses on developing innovative solutions for sustainability and diversity management in the fashion industry (Obaid, 2021a). The underlying aim of the Saudi Fashion Commission is "to enable the development of a thriving Saudi fashion industry, sustainable and inclusive, fully integrated along the value chain, maximizing local talent, experiences, and competencies" (Fashion Commission, 2021b).

The country's commitment and confidence towards attaining the aforementioned things are also reflected in the words of Princess Reema Bandar Al-Saud, Saudi Arabia's ambassador to the US, who had proclaimed that Saudi Arabia would be a leader in fashion in the future. The sentiments are matched in the words of Burak Cakmak, CEO of the Fashion Commission, when he said:

> Saudi Arabia can serve as an example of how to build an innovative, sustainable and appropriate fashion sector, locally and internationally. By working with innovators in the sector, attracting retail experiences and establishing partnerships for education, business development and entrepreneurship, he added, the Kingdom will be able to develop the processes and brands of local businesses to improve them in line with international best practices. (Alkhudair, 2021, p. 2053)

Therefore, it can be observed that the Saudi government is committed to developing the sustainable fashion sector in the country and is focused on creating entrepreneurship support.

6.2.4 Female Entrepreneurship in Luxury Fashion Industry in Saudi Arabia

The luxury fashion industry in Saudi Arabia has seen the emergence of numerous female entrepreneurs that have been doing a commendable job (Basahal, 2020). However, as the luxury fashion industry has expanded, there have been several challenges that these female entrepreneurs face, such as the need to become a part of the circular economy paradigm initiated by the Saudi government (Aboumoghli & Alabdallah, 2019). While promoted and encouraged by the government, ensuring sustainability alignment largely falls on the business owners, as society holds them accountable for degrading the environment for their profit. Under pressure from society and the consumers, entrepreneurs have focused on developing more sustainable production pipelines and adopting practices, such as re-looping or redirecting unused garments towards new supply chains and recycling processes for garments (AlSabban & Issa, 2020). However, there is also an emerging trend among entrepreneurs, driven by their values and ethics (Li & Leonas, 2019), which led to the emergence of female luxury fashion micro-entrepreneurs who deal in second-hand luxury fashion garments (Hu et al., 2019). It is also true in the context of Saudi luxury fashion entrepreneurs (Aljuwaiber, 2020), who have overcome substantial socio-cultural and educational barriers to succeed in business (Abdulghaffar & Akkad, 2021).

While on the one hand, these recycled luxury fashion garments or textiles are made from renewable sources and reduce the industry's carbon footprint, on the other hand, they compete with the traditional luxury fashion industry (Athwal et al., 2019). In Western countries, the second-hand consumption of luxury goods is already de-stigmatised (Battle et al., 2018), and the second-hand or vintage fashion markets are well-accepted (Barnes & Lea-Greenwood, 2018). However, in the context of Saudi Arabia, recycled or pre-owned garments, or even garments made by using innovative processes such as renewable sources of material, may not necessarily be seen as acceptable.

In the Saudi context, through *zakat* (charity), second-hand is accepted and recycled and passed on to those in need (AlSabban et al., 2014). In addition, charitable associations created in Saudi have launched initiatives to help upcycle these unwanted items to help to promote a more sustainable environment (Hamdan, 2021a). It is also true in businesses, where left-over materials from production processes can be donated to charities, thus simulating a circular economy (AlSabban et al., 2014; Elshaer et al., 2021). Each year, Saudi Arabia discards tonnes of textiles and clothing in its recycling centres and landfills (Sakshi, 2021). To promote a sustainable environment, charitable organisations in Saudi Arabia have initiated programmes to assist in the recycling of these discarded items (Sakshi, 2021). Consumers from the Middle East generally prefer purchasing new goods (AETOSWire, 2021), and second-hand clothes are perceived as being for charity, and people do not sell or buy them (Alfawaz, 2014).

New luxury has always been popular with consumers across the Middle East to buy, although used goods are a new issue in the region (Assomull, 2021). Currently, online luxury in the Middle East is expanding. Farfetch, an online luxury retailer, wants to combine both by extending Farfetch Second Life, its handbag resale service, to the United Arab Emirates, Kuwait, and Saudi Arabia (Assomull, 2021). The reason for this is that younger and more environmentally conscious consumers want to buy high-end goods that are in good condition (AETOSWire, 2021; Ramadan & Nsouli, 2021). With the help of social media and the media, eco-fashion and sustainable trade values have played a vital role in promoting this development to the new generation of Saudis. Moreover, the current economic climate has encouraged a change in consumer opinions concerning wearing and utilising second-hand goods combined with vintage inspirations in current fashion designs (Obaid, 2021b).

The Middle East's passion for luxury stimulates rapid growth in the second-hand luxury goods sector. Whilst the pandemic has caused reduced in-person shopping, online platforms have benefitted from a significant increase in demand from millennials eager to purchase high-end goods at affordable prices (AETOSWire, 2021). More recently, in Saudi Arabia, there has been a shift in perceptions around pre-owned goods and positive acceptance of resale or recycled products (Obaid, 2021b). This shift is primarily attributed to the growth of Generation Z (Gen Z) and Millennials (Gen Y), who are known to prefer ethical, eco-friendly, and sustainable consumption (Kim et al., 2021a). These generations favour the second-hand luxury fashion market in Saudi Arabia (AETOSWire, 2021; Obaid, 2021b).

Moreover, with the expected rise in the country's population and the predominant percentage of young people in the population, second-hand luxury fashion can be expected to grow (Obaid, 2021b). However, second-hand luxury fashion forms just one aspect of the circular economy, as seen in the previous section (Centobelli et al., 2020a). Entrepreneurs need to develop more innovative approaches to ensuring that their production, design, and processes are leading to environmental sustainability and that they are focused on long-term usage of the clothes. Thus, entrepreneurs need to be vigilant about the latest technologies that can enable them to reuse material innovatively, repurpose and redesign clothes, and develop outputs that are appealing and environmentally friendly (Santasalo-Aarnio et al., 2017). According to Veleva and Bodkin (2018), knowledge and awareness around the circular economy and the strategies and methodologies to adapt to ensure that they participate in and contribute to the circular economy are essential for luxury business (Centobelli et al., 2020b). It is here that the role of the government is highlighted, as the government is able to provide training and development for entrepreneurs (Welsh et al., 2014) in the circular economy, as well as to give them hands-on support to implement the principles of circular economy (Veleva & Bodkin, 2018). It also needs to renew its knowledge base and practices and harness new information on recycling technologies (Niinimäki & Karell, 2020).

6.3 Methodology

This research employed a qualitative methodology (Guba & Lincoln 1994; Symon & Cassell, 2012) to gain insights into female entrepreneurs and their experiences with the circular economy in Saudi Arabia. A sample of 22 Saudi female luxury fashion entrepreneurs were purposively selected (Table 6.1). A purposive sample was more suitable as it included participants who were likely to have insights and knowledge and were interested in discussing their problems and initiatives. Prior to conducting any research, ethical approval was obtained. Each interview lasted 60 minutes and was semi-structured.

Examples of the probing questions asked are as follows:

- How does the Saudi vision 2030 influence your business?
- What challenges have you faced in the past, if any?
- Are there any challenges you currently face in operating your business?

The interviews have been transcribed verbatim and carefully translated, ensuring the process is reliable and preserves the data's integrity (Al-Amer et al., 2016). Data were analysed using a grounded-analysis approach of Easterby-Smith et al.'s (2018) seven-step guide, involving

Table 6.1 Summary of participants

No.	Luxury fashion categories	Duration of interview
E1	Evening apparel	68.51
E2	Evening apparel and casual	56.43
E3	Ready-to-wear and evening apparel	59.15
E4	Evening apparel	65.33
E5	Evening apparel	53.42
E6	Evening apparel and casual	58.39
E7	Ready-to-wear and evening apparel	47.05
E8	Children, casual, weddings, and evening apparel	78.34
E9	Ready-to-wear and abaya	49.15
E10	Ready-to-wear and casual	39.07
E11	Ready-to-wear and evening apparel	51.50
E12	Ready-to-wear, sustainable fashion, and accessories, apparel	83.55
E13	Ready-to-wear, sustainable fashion, and accessories, apparel	59.28
E14	Ready-to-wear, sustainable fashion, and accessories, apparel	89.54
E15	Ready-to-wear and evening apparel	58.12
E16	Ready-to-wear and abaya	57.46
E17	Sustainable fashion and modern folk garment	89.55
E18	Ready-to-wear and evening apparel	40.25
E19	Ethical and modern national garment-abaya	55.48
E20	Bags, accessories, apparel	49.58
E21	Ethical and modern national garment-abaya	39.42
E22	Designer luxury womenswear fashion (bags, accessories, apparel)	44.54

familiarisation, reflection, open-coding, conceptualisation, focused re-coding, linking, and re-evaluation. It is important to note that grounded analysis is open to discoveries within the data. To ensure coherence, clarity, and continuity, the researchers examined the data independently.

6.4 Findings and Discussion

6.4.1 Understanding Concepts of Circular Economy and Sustainability

This chapter seeks to better understand how Saudi female luxury fashion entrepreneurs think about sustainability and the circular economy. The analysis of the interviews revealed that the participants had varying ideas about sustainability and circular economy techniques. Only a handful of attendees were familiar with the new circular economy strategy and how it may help the Saudi sustainable fashion industry's future. As stated by E17, "early 2020, I met with one of my employees at the Saudi Ministry of Culture to explore the circumstances of manufacturers in Saudi to help the Kingdom fulfil its Vision 2030 goals… in sustainability". These findings indicate that the participants are receptive to the idea of exploring how they can attain sustainability and are open to receiving information and support to guide them through their endeavour. This is also an important finding because it shows that the participants regarded the Saudi Vision 2030 as vital and desired to implement the circular economy strategy to achieve their sustainability goals.

However, the findings also indicate that the current level of understanding is limited for the concepts and constructs related to the circular economy and the strategies that they can implement to participate in the circular economy. Although the Ministry of Culture offers a variety of sustainability and circular economy training programmes (Arab News, 2021a; Fashion Commission, 2021a), female entrepreneurs are unfamiliar with the programmes due to a lack of effective communication and engagement from the Commission. For example, one of the participants, E3, said, "We need some courses to understand what sustainable means". Also, E6 stated, "we do not have enough awareness of the new plan. We need a link between the Ministry of Culture and us". It demonstrates a gap between female entrepreneurs and the organisers of circular economy programmes. E1, E2, E18, E21, and E22 also felt that they do not have sufficient knowledge or need a more direct connection and knowledge transfer from the Ministry of Culture to support them in their business.

These findings show that the respondents desire more hands-on guidance from the Ministry of Culture to help them understand and implement the circular economy concepts. They may even want greater clarity on the essential meaning of the terms like sustainability and circular economy. Therefore, training is required by establishing new knowledge about sustainable methods to adapt their practices to the new circular economy setting (Welsh et al., 2014; Dan & Østergaard, 2021). The E10 stated, "The implementation of Saudi Vision 2030 in the Saudi fashion industry has had limited success because It is important to raise fashion entrepreneurs' awareness of the benefits of the circular economy so that they may better manage their businesses". The findings reveal that the respondents believed that understanding sustainability concerning circular economy practices in their field is essential to achieving the Saudi Vision 2030.

The aforementioned theme is relevant and needs to be contextualised within the existing literature. As seen from the review of literature, the field of circular economy is emerging with many different theories and practical solutions being put forward (Santasalo-Aarnio et al., 2017). This means that there is a need for structuring the available knowledge and interpreting it for the entrepreneurs who can undertake practical implementation of the circular economy strategies without the need for deep diving into theory and research (Veleva & Bodkin, 2018). In the case of Saudi female entrepreneurs, the confusion and lack of direction was apparent from the interviews, and it indicated an unfulfilled need in the country. The participants appeared to be enthusiastic and passionate about their businesses as well as environmental protection, but they lacked specific understanding about how to conduct their businesses in an environmentally sustainable manner. There is therefore a need to provide them with effective communications and guidance on the subject.

6.4.2 Generation Y's Evolving Conception and the Circular Economy of Luxury Fashion in Saudi Arabia

Although the findings' underscored challenges were related to the sociocultural barriers and stigma against using recycled products, especially clothing. It was revealed in the interviews by several participants, who mentioned that it is not considered acceptable in Saudi society to reuse material that has been classified as waste. For example, E4 mentioned that "society's culture is sometimes not accepting recycled or reused material, even the recycled raw material", which probably meant that the Saudi society and culture opposed using recycled products.

It was an important finding, as it underscores the cultural barriers that luxury fashion entrepreneurs may be facing in the country. However, "the new generation is poised to push us toward greater inventiveness because they are more knowledgeable on sustainability" (E11). It is also imperative that fashion brands understand how Generation Y perceives their products and their impact on the bottom line. To successfully address consumer needs, fashion business owners need to understand their attitudes on circular fashion (Kim et al., 2021a). The study results confirmed that luxury fashion entrepreneurs indicate that Luxury fashion brands need to find innovative ways of communicating their sustainability business model to their consumers. "We need to be more creative to attract the new Generation" (E11). However, they currently lack a fuller understanding of the practical implementation of these strategies in their business processes.

Moreover, the interviews showed that new generations (the millennials and Generation Z) of entrepreneurs try to improve our business and seek international fashion standards and a desire to learn from the various resources available such as online. These findings imply that Saudi female entrepreneurs in the luxury fashion industry endeavoured to employ circular economy practices and perceived sustainability as vital to their business from resources. Additionally, they have assisted female Gen Y entrepreneurs to shape their collective online community: "We are a collection of entrepreneurs that constantly post and share comparable knowledge regarding circular economy to achieve the Saudi vision 2030" (E9). Study participants confirm the new generation (millennials, Gen Z) of Saudi entrepreneurs are more open to learning and developing in their approach to sustainability in a way to meet the Saudi vision 2030. "I believe that the majority of Saudi fashion entrepreneurs are from the younger generation, the millennials and Generation Z, and they will do a great deal of work demonstrating their approach to the circular economy if they have sufficient understanding about the Saudi Vision 2030" (E10). This theme explains that when luxury fashion industry transitions to a more sustainable business strategy, Gen Y/Gen Z entrepreneurs value and appreciate sustainable luxury. It also indicates gaps at the policy level, as the government does not include any specific programmes or directions for overcoming this cultural barrier. However, governments and politicians can remove existing impediments at the market, society, and innovation system and promote the implementation of targeted initiatives because their actions can function as drivers or hurdles to companies' move to a circular economy (Centobelli et al., 2020a). It also indicates that female entrepreneurs are eager to seek information and comprehensive support to enable and empower them to achieve their sustainability goals.

Additionally, the findings can also be interpreted to mean that the entrepreneurs, though eager and receptive to sustainability, are unable to find and take advantage of resources or education around sustainable business operations. These findings also highlight that the participants were vague on sustainability and how they can successfully implement circular economy strategies in their business strategies. Entrepreneurship is critical for advancing circular luxury on a sustainable basis. These entrepreneurs are transformational leaders who work to improve people's lives and the environment (Gardetti & Torres, 2013). Nevertheless, the findings also indicate the eagerness to learn and a willingness to obtain more information and guidance.

6.4.3 The Beginning of Recycling Operations

Although it was seen from the interview analysis that despite the limited understanding of policy or information availability, some participants were using the available capabilities and available practices related to sustainability; for example, some participants had revealed that they could reuse old clothes and redesign and refurbish them into new dresses. However, the main challenge was the availability and supply of raw materials that could support their sustainability designs and marketing approaches. For example, participant E19 mentioned:

We are looking to start to recycle our product by asking the customer to bring the old product to get money off coupons and then sell this product as second-hand as a part of the realisation of (Saudi)Vision 2030 … the revenues of the sale could be donated to charitable organisations.

By drawing on their past knowledge of culture, participants strive to execute a simple idea in that second-hand links to the circular economy and Saudi Vision 2030. Thus, rather than throwing clothes away, they should donate them to charities (AlSabban et al., 2014; Elshaer et al., 2021).

This is nevertheless a strategy already documented in literature where companies develop a subscription-based collection where people can subscribe to the company and give their old clothes to it, in return for which they are given coupons that they can redeem on their future purchases (Hvass & Pedersen, 2019; Dissanayake & Weerasinghe, 2022). Similarly, E8 mentioned, "I had a policy for our customers to return the old dresses from our brand and re-design them again with a new cut and affordable prices". However, this finding underscores the fact that such redesigning could be limited in scope as the left-over pieces used in the redesigning could be scarce, and hence the overall number of dresses that can be made may be limited. Participants attempt to implement a sample idea of recycling that relates to the circular economy and Saudi Vision 2030 by drawing on their prior knowledge of recycling. It permits boosting product reuse and recycling (Niinimäki & Karell, 2020; Dissanayake & Weerasinghe, 2022). It is evident that by providing incentives to customers, the entrepreneurs were creating motivation for recycling old clothes and keeping them further in active use. It also displays creativity in managing waste and shows the entrepreneurs' ingenuity in re-marketing the used garments and putting them back in the loop.

6.4.4 Lack of Circular Resource Loops

E24 claimed, "I try to select eco-friendly materials for design". The fact that Saudi female fashion entrepreneurs started such a strategy without getting any direct information or support from the government is commendable. However, it also underscores the need for the policymakers to compile a list of best practices and grass-root level strategies that similar other entrepreneurs can implement. Similarly, other participants also mentioned indulging in innovative strategies for sustainability and revealed the challenges they faced in doing so. For example, some of the participants, predominantly E12, mentioned that: "as a business usually try to be more sustainable. Although the limited resources use the leftovers fabric to design the patchwork (sewing together pieces of fabric) as reused for new design". It was also reiterated by participants E14 and E17. The challenge here was the limited availability of reusable raw material, as though the entrepreneurs were attempting to use the innovative design strategy to ensure sustainability, the supply of their raw materials limited them.

Additionally, a related challenge was the availability of reusable and innovative fabrics for raw materials, as the entrepreneurs did not seem to have access to these. Some respondents mentioned that they were hesitant about using recycled processes or materials, as they believed that using recycled fibres to improve sustainability will limit resources. According to E14, "In Saudi Arabia, we do not have natural resources for the textile industry, but with innovations in recycled and sustainable sources, this is something we can do. However, these may restrict our fabric usage". These findings indicate that there may be a need for investing in research and development so that more innovative and reusable raw materials can be developed that can be used effectively in the luxury fashion industry. Similarly, a challenge related to lack of resources was revealed.

Moreover, a challenge related to lack of resources was revealed in the form of lack of trained Saudi artisans, as E13 mentioned: "We need as entrepreneurs to empower Saudi artisans who are a part of the circular economy for Saudi fashion in our product and pass a message to the world for the future need of sustainability in fashion". The aforementioned findings again point out a gap that needs to be addressed by the government and policy-making agencies, so that the supply chains and skilled and trained artisans can be made available for the Saudi female entrepreneurs to successfully implement innovative strategies for the circular economy.

The shift to the circular economy of people possesses the required competencies and knowledge of circular design (Ellen MacArthur Foundation, 2020, 2021). These findings are also important as they point out to the fact that the respondents have some tentative understanding of circular economy (which includes using innovative processes or raw material, as the patchwork that one of the participants suggested). They may not have full access to resources or knowledge that would assist them in efficiently implementing it, but they endeavour to complete the loop as a sort of sustainable development in which the objective is to transform an unsustainable situation into a sustainable one (Niinimäki & Karell, 2020). It implies that there is a need for improving access to environmentally friendly raw materials and resources for the fashion luxury entrepreneurs so that they can make the best use of circular economy resources.

6.5 Conclusion

The findings of this study answer the Ellen MacArthur Foundation's (2018) report on the circular fashion industry and reinforce the requirement of the luxury fashion industry to incorporate reparable and long-lasting design elements. The research's findings indicated that the female entrepreneurs in the fashion luxury market faced several challenges, prominent being a lack of clarity on the concepts like sustainability and circular economy. Other challenges they face include lack of availability of innovative raw materials or reusable raw material, lack of trained artisans, lack of knowledge around specific circular economy strategies, and lack of training and support for them as sustainability entrepreneurs. Additionally, despite the vague understanding of sustainability, these female entrepreneurs face socio-cultural barriers as Saudi society does not accept recycled or made-from-waste products. The Saudi Vision 2030 and the Circular Economy Strategy provide some information and guidance, but they seem inadequate. Policymakers need to develop more explicit and comprehensive strategies that female entrepreneurs can easily understand and interpret. There is also a need for policy changes to ensure that Saudi society becomes more accepting of the concepts of recycling and reusing fabric and garments. The expectations of the fashion entrepreneur are high in terms of responsibilities and based on the interviews; the entrepreneur has the possibility for increased impact. However, being granted authority and mandate by management and being accepted by the entrepreneur is contingent on several factors, including management's willingness to train and then listen to designers; the entrepreneur's personality and the breadth of his/her knowledge bank regarding circular strategies, as well as enthusiasm to take on new responsibilities; and the general context and a more entrepreneurial mindset. Along with these elements, the unique organisational setting and structure can be considered significant.

This research fills a gap in understanding how entrepreneurs handle circular economy strategies (Centobelli et al., 2020a). Most research around fashion take-back programmes have focused on consumers' perceptions and participation to support take-back programmes (Balconi et al., 2019; Assomull, 2021). However, according to the researcher knowledge, no research in the Middle East has explored take-back programmes from an entrepreneur's perspective, nor has it shown how the entrepreneurs manage the circular economy strategies.

6.6 Limitations of the Research and Future Developments

The research suffered from some limitations that can be overcome in future research. For example, the current research used a purposive sample, which could have limited the number and diversity of participants. Additionally, the research relied solely on qualitative methods, which, while providing rich and contextual data, also suffered from the limitation that only a smaller sample size could be included. Therefore, a mixed methodology is recommended for future research to include larger samples. Further, the research focused only on Saudi Arabia, and any future research may benefit by adopting a comparative approach where other countries in the region can be explored. Finally, the research was focused on the perceptions of the entrepreneurs alone, and future research needs to include the consumers' perceptions to present a more holistic understanding of the issue. Moreover, for future research, a more incredible amount of investigation is needed to build long-term business plans for a new environment for sustainable fashion in Saudi Arabia.

6.7 Practitioner Implications

Sustainable fashion is a collaborative effort involving industry sectors, consumers, and governments. All stakeholders must work cooperatively to effect change. They are committed to the circular economy and the development of restorative economic concepts (Ellen MacArthur Foundation, 2012). The fashion industry has demonstrated resilience by adopting more sustainable business practices; it may use textile recycling innovations to withstand a radical system transformation. Saudi female entrepreneurs apply the circular economy approach in the luxury fashion industry to highlight the importance and power of female entrepreneurs in making fashion circular. Identifying such key actors and showing how they operate is a starting point for change and essential to moving the fashion industry towards a new circular textile economy (Ellen MacArthur Foundation, 2017). This research shows that implementing circularity can increase business revenue. Moreover, the need to develop new products and methods of production is immanent, in that it can aid in attracting new customers such as Generation Y.

The study also encourages practitioners to explore technology that can enable them to develop and design renewable and reusable products and stay longer in use. By providing insights about the importance of collaborative effort and systemic change, the study encourages practitioners to develop networks and processes that enable them to align with the circular economy principles.

References

Abdulghaffar, N. A., & Akkad, G. S. (2021). Internal and external barriers to entrepreneurship in Saudi Arabia. *Digest of Middle East Studies, 30*(2), 116–134. https://doi.org/10.1111/dome.12231.

Aboumoghli, A., & Alabdallah, G. M. (2019). A systematic review of women entrepreneurs opportunities and challenges in Saudi Arabia. *Journal of Entrepreneurship Education, 22*(6), 1–14.

AETOSWire. (2021). Millennials propel rapid growth of the second-hand luxury goods market in the Middle East 26 May. Accessed October 21, 2021, from https://alrasub.com/millennials-propel-rapid-growth-of-the-second-hand-luxury-goods-market-in-the-middle-east/

Al-Amer, R., Ramjan, L., Glew, P., Darwish, M., & Salamonson, Y. (2016). Language Translation Challenges with Arabic Speakers Participating In Qualitative Research Studies. *International Journal Of Nursing Studies, 54*, 150–157.

Alfawaz, N. (2014). Second-hand clothes not an option for Saudis. 15 November. Accessed October 21, 2021, from https://www.arabnews.com/saudi-arabia/news/660336

Aljuwaiber, A. (2020). Entrepreneurship research in the Middle East and North Africa: Trends, challenges, and sustainability issues. *Journal of Entrepreneurship in Emerging Economies., 13*(3), 380–426. https://doi.org/10.1108/JEEE-08-2019-0123

Alkhudair, D. (2021, June 12). New Saudi fashion forum to launch next week with live event in New York and Riyadh. *Arab News.* Accessed June 21, 2021, from https://www.arabnews.com/node/1875081/saudi-arabia

Almulhim, A. I., & Abubakar, I. R. (2021). Understanding Public Environmental Awareness and Attitudes Toward Circular Economy Transition in Saudi Arabia. *Sustainability, 13*(18), 10157.

AlSabban, N., Al Sabban, Y., & Rahatullah, M. K. (2014). Exploring corporate social responsibility policies in family owned businesses of Saudi Arabia. *International Journal of Research Studies in Management, 3*(2), 51–58.

AlSabban, L., & Issa, T. (2020). Sustainability awareness in Saudi Arabia. In T. Issa, T. Issa, T. B. Issa, & P. Isaias (Eds.), *Sustainability awareness and green information technologies. Green energy and technology* (pp. 101–131). Springer.

Arab News. (2021a). KSA fashion commission backs luxury designs with 100 Saudi brands programme. 04 June. Accessed June 20, 2021, from https://www.arabnews.com/node/1870166/saudi-arabia

Arab News. (2021b, October 8). Saudi Arabia's Fashion Commission to host Fashion Futures event in December. *Arab News*. Retrieved January 20, 2022, from https://Www.Arabnews.Com/Node/1944096/Saudi-Arabia

Assomull, S. (2021). Why Farfetch is bringing its handbag resale business to the Middle East13 October voguebusiness. Accessed October 23, 2021, from https://www.voguebusiness.com/consumers/why-farfetch-is-bringing-its-handbag-resale-business-to-the-middle-east

Athwal, N., Wells, V. K., Carrigan, M., & Henninger, C. E. (2019). Sustainable luxury marketing: A synthesis and research agenda. *International Journal of Management Reviews, 21*(4), 405–426.

Balconi, M., Sebastiani, R., & Angioletti, L. (2019). A neuroscientific approach to explore consumers' intentions towards sustainability within the luxury fashion industry. *Sustainability, 11*(18), 5105. https://doi.org/10.3390/su11185105

Barnes, L., & Lea-Greenwood, G. (2018). Pre-loved? Analysing the Dubai luxe resale market. In *Vintage luxury fashion* (pp. 63–78). Cham: Palgrave Macmillan.

Basahal, A. S. (2020). Female entrepreneurship in Saudi Arabia: Motivations and barriers. *International Journal of Human Potentials Management, 2*(2), 1–17.

Battle, A., Ryding, D., & Henninger, C. E. (2018). Access-based consumption: A new business model for luxury and second-hand fashion business?. In *Vintage luxury fashion* (pp. 29–44). Palgrave Macmillan.

Bertassini, A. C., Zanon, L. G., Azarias, J. G., Gerolamo, M. C., & Ometto, A. R. (2021). Circular business ecosystem innovation: A guide for mapping stakeholders, capturing values and finding new opportunities. *Sustainable Production and Consumption, 27*, 436–448.

Bocken, N. M., De Pauw, I., Bakker, C., & Van Der Grinten, B. (2016). Product design and business model strategies for a circular economy. *Journal of Industrial and Production Engineering, 33*(5), 308–320.

Brydges, T. (2021). Closing the loop on take, make, waste: Investigating circular economy practices in the Swedish fashion industry. *Journal of Cleaner Production, 293*, 126245.

Cabigiosu, A. (2020). An overview of the luxury fashion industry. *Digitalisation in the Luxury Fashion Industry*, 9–31.

Centobelli, P., Cerchione, R., Chiaroni, D., Del Vecchio, P., & Urbinati, A. (2020a). Designing business models in circular economy: A systematic literature review and research agenda. *Business Strategy and the Environment, 29*(4), 1734–1749. https://doi.org/10.1002/bse.2466

Centobelli, P., Cerchione, R., & Mittal, A. (2020b). Managing sustainability in luxury industry to pursue circular economy strategies. *Business Strategy and the Environment, 30*(1), 432–462. https://doi.org/10.1002/bse.2630

Dan, M. C., & Østergaard, T. (2021). Circular Fashion: The New Roles of Designers In Organisations Transitioning To A Circular Economy. *The Design Journal, 24*(6), 1001–1021.

Dissanayake, D. G. K., & Weerasinghe, D. (2022). Towards circular economy in fashion: Review of strategies, barriers and enablers. *Circular Economy and Sustainability, 2*, 25–45.

Easterby-Smith, M., Thorpe, R., & Jackson, P. (2018). *Management research* (5th ed.). SAGE Publications.

Ellen MacArthur Foundation. (2012). *Towards the circular economy: Economic and business rationale for an accelerated transition. Isle of Wight*. Ellen Macarthur Foundation. Retrieved January 11, 2022, from https://www.ellenmacarthurfoundation.org/assets/downloads/publications/Ellen-Macarthur-Foundation-Towards-The-Circular-Economy-Vol.1.Pdf

Ellen MacArthur Foundation. (2017). *What is the circular economy?* EMF (online). Retrieved March 7, 2021, from https://www.ellenmacarthurfoundation.org/circular-economy/what-is-the-circular-economy

Ellen MacArthur Foundation. (2018). *The circular economy in detail*. Ellen Macarthur Foundation. Retrieved January 19, 2022, from https://archive.ellenmacarthurfoundation.org/explore/The-Circular-Economy-In-Detail

Ellen MacArthur Foundation. (2020). *Fashion and the circular economy*. Ellen Macarthur Foundation. Retrieved January 14, 2022, from https://archive.ellenmacarthurfoundation.org/explore/Fashion-And-The-Circular-Economy

Ellen MacArthur Foundation. (2021). *Circular design for fashion foundation*. Ellen Macarthur Foundation.

Elshaer, I., Sobaih, A. E. E., Alyahya, M., & Abu Elnasr, A. (2021). The impact of religiosity and food consumption culture on food waste intention in Saudi Arabia. *Sustainability, 13*(11), 6473.

Fashion Commission. (2021a). SAUDI 100 BRANDS 360° Professional Development Programme for Saudi Arabian Luxury and Fashion Brands. Accessed June 20, 2021, from https://saudi100brands.com/#community

Fashion Commission. (2021b). Fashion Commission. Accessed June 20, 2021, from https://fashion.moc.cards/about-fashion-commission-final

Gardetti, M. A., & Torres, A. L. (2013). Entrepreneurship innovation and luxury. *Journal of Corporate Citizenship, 52*, 55–75.

Geissdoerfer, M., Savaget, P., Bocken, N. M., & Hultink, E. J. (2017). The circular economy–a new sustainability paradigm? *Journal of Cleaner Production, 143*, 757–768.

Guba, E. G., & Lincoln, Y. S. (1994). Competing paradigms in qualitative research. *Handbook of qualitative research, 2*(163–194), 105.

Hamdan, S. (2021a). Saudi charities create sustainable friendly fashion October 11. Accessed October 26, 2021, from https://www.arabnews.com/node/1945376/saudi-arabia

Hamdan, S. (2021b). Sustainable fashion: What will we wear in the future?. October 26. Accessed October 26, 2021, from https://www.arabnews.com/node/1955201

Henninger, C. E. (2021a). Creative marketing and the clothes swapping phenomenon. In *Creativity and marketing: The fuel for success*. Emerald Publishing Limited.

Henninger, C. E., Brydges, T., Iran, S., & Vladimirova, K. (2021b). Collaborative fashion consumption–A synthesis and future research agenda. *Journal of Cleaner Production, 128648*.

Henninger, C., Jones, C., Boardman, R., & McCormick, H. (2018). Collaborative Consumption and the Fashion Industry. In *Sustainable Fashion in a Circular Economy*. https://aaltodoc.aalto.fi/bitstream/handle/123456789/36608/isbn9789526000909.pdf?sequence=1&isAllowed=y

Henninger, C. E., Alevizou, P. J., Goworek, H., & Ryding, D. (Eds.). (2017). *Sustainability in fashion: A cradle to upcycle approach*. Springer.

Henninger, C. E., Alevizou, P. J., & Oates, C. J. (2016). What is sustainable fashion? *Journal of Fashion Marketing and Management, 20*(4), 400–416.

Henninger, C. E., Blazquez Cano, M., Boardman, R., Jones, C., McCormick, H., & Sahab, S. (2020). Cradle-to-cradle vs consumer preferences. In I. Choudhury & S. Hashmi (Eds.).

Hidalgo, D., Martín-Marroquín, J. M., & Corona, F. (2019). A multi-waste management concept as a basis towards a circular economy model. *Renewable and Sustainable Energy Reviews, 111*, 481–489.

Hu, S., Henninger, C.E., Boardman, R., & Ryding, D. (2019). Challenging current fashion business models: Entrepreneurship through access-based consumption in the second-hand luxury garment sector within a circular economy. In *Sustainable luxury* (pp. 39–54). Springer.

Hvass, K. K., & Pedersen, E. R. G. (2019). Toward circular economy of fashion: Experiences from a brand's product take-back initiative. *Journal of Fashion Marketing and Management: An International Journal, 23*(9), 345–365.

Jain, S., & Mishra, S. (2019). Sadhu—On the pathway of luxury sustainable circular value model. In *Sustainable luxury* (pp. 55–82). Springer.

Ki, C. W., Chong, S. M., & Ha-Brookshire, J. E. (2020). How fashion can achieve sustainable development through a circular economy and stakeholder engagement: A systematic literature review. *Corporate Social Responsibility and Environmental Management, 27*(6), 2401–2424.

Kim, I., Jung, H. J., & Lee, Y. (2021a). Consumers' value and risk perceptions of circular fashion: Comparison between second-hand, upcycled, and recycled clothing. *Sustainability, 13*(3), 1208.

Kim, E., Fiore, A. M., Payne, A., and Kim, H. (2021b). *Fashion Trends: Analysis and Forecasting*. Bloomsbury Publishing.

Li, J., & Leonas, K. K. (2019). Trends of sustainable development among the luxury industry. In *Sustainable luxury* (pp. 107–126). Springer.

Malek, C. (2021, September 23). How circular carbon economy provides a framework for a sustainable future. *Arab News*. Retrieved January 20, 2022, from https://Www.Arabnews.Com/Node/1934466/Middle-East

Mishra, S., Jain, S., and Jham, V. (2021). Luxury rental purchase intention among millennials: A cross-national study. *Thunderbird International Business Review, 63*(4), 503–516.

Mishra, S., Jain, S., & Malhotra, G. (2020). The anatomy of the circular economy transition in the fashion industry. *Social Responsibility Journal, 17*(4), 524–542.

Moorhouse, D., & Moorhouse, D. (2017). Sustainable design: Circular economy in fashion and textiles. *The Design Journal, 20*(Suppl 1), S1948–S1959.

Murray, A., Skene, K., & Haynes, K. (2017). The circular economy: An interdisciplinary exploration of the concept and application in a global context. *Journal of Business Ethics, 140*(3), 369–380.

Niinimäki, K., & Karell, E. (2020). Closing the loop: Intentional fashion design defined by recycling technologies. In *Technology-driven sustainability* (pp. 7–25). Palgrave Macmillan.

Niinimäki, K. (2017). Fashion in a circular economy. In *Sustainability in fashion* (pp. 151–169). Palgrave Macmillan.

Obaid, R. (2021b). Vintage items given new lease of life by Saudi Gen Z admirers. *Arab News*, March 14. Accessed October 21, 2021, from https://www.arabnews.com/node/1825051/saudi-arabia

Obaid, R. (2021a, June 19). *Driving the future of Saudi fashion toward sustainability, diversity, innovation*. Accessed June 21, 2021, from https://www.arabnews.com/node/1879196/saudi-arabia

Ræbild, U., & Bang, A. L. (2017). Rethinking the fashion collection as a design strategic tool in a circular economy. *The Design Journal*, 20(Suppl 1), S589–S599.
Ramadan, Z., & Mrad, M. (2017). Fashionable stereotypes and evolving trends in the United Arab Emirates. *Customer Needs and Solutions*, 4(1), 28–36.
Ramadan, Z., & Nsouli, N. Z. (2021). Luxury fashion start-up brands' digital strategies with female Gen Y in the Middle East. *Journal of Fashion Marketing and Management*. https://doi.org/10.1108/JFMM-10-2020-0222.
Sakshi, K. S. (2021, October 11). Saudi Charitable Organisations endorse vision 2030 sustainability goals through recycled clothing. Retrieved January 22, 2022, from https://Www.Theglobaleconomics.Com/2021/10/11/Saudi-Charitable-Organisations/#:~:Text=The%20Kingdom%20of%20Saudi%20Arabia,To%20endorse%20a%20sustainable%20environment
Santasalo-Aarnio, A., Hänninen, A., & Serna-Guerrero, R. (2017). Circular economy design forum-introducing entrepreneurial mindset and circularity to teaching. In *Annual Conference of the European Society for Engineering Education* (pp. 104–111). SEFI Société Européenne pour la Formation des Ingénieurs.
Shirvanimoghaddam, K., Motamed, B., Ramakrishna, S., & Naebe, M. (2020). Death by waste: Fashion and textile circular economy case. *Science of the Total Environment*, 718, 137317.
Symon, G., & Cassell, C. (2012). Assessing qualitative research. In G. Symon & C. Cassell (Eds.), *Organisational qualitative research: Core methods and current challenges*. Sage.
Veleva, V., & Bodkin, G. (2018). Corporate-entrepreneur collaborations to advance a circular economy. *Journal of Cleaner Production*, 188, 20–37.
Welsh, D. H., Memili, E., Kaciak, E., & Al Sadoon, A. (2014). Saudi women entrepreneurs: A growing economic segment. *Journal of Business Research*, 67(5), 758–762.

7

Sustainable Supply Chain Process of the Luxury *Kente* Textile: Introducing Heritage into the Sustainability Framework

Sharon Nunoo

7.1 Introduction: Setting the Scene

The twenty-first century sees the rise of the conscious consumer, who is concerned about environmental and social impacts their purchase decisions may have on their natural surroundings (Rickenbacher, 2020). The phrase *we are what we consume* (Belk, 1988) gains renewed interest, with consumers seeking to portray an image that proclaims this environmental and social responsibility, especially within the luxury context (Rickenbacher, 2020; Roberts, 2020). The latter sees five key consumer trends emerging: (1) *Blingtastics*, who seek to show off luxury wear; (2) *Exclusivists*, who hunt for special editions at a high price; (3) *Old Money*, who seek value brand heritage and authenticity; (4) *New Ascetics*, who focus on artisan and local products and are drawn towards more green luxury; and (5) *Novelty Junkies*, who are highly influenced by new

S. Nunoo (✉)
Department of Materials, The University of Manchester, Manchester, UK
e-mail: Sharon.Nunoo@manchester.ac.uk

technology and look for new trends (Roberts, 2020). Within this chapter, *Old Money* and *New Ascetics* consumers are of particular interest, as these are consumers that are more actively pursuing the sustainable trend, thereby also taking heritage into consideration. These consumer types will be explored further in the latter part of this chapter.

Research on luxury consumption is not new per se, but has gained momentum in the past decade within the fashion and jewellery context (e.g. Ryding et al., 2018; Athwal et al., 2019). Jewellery is often associated with precious material necklaces, bracelets, rings, or earrings that people use to adorn and differentiate themselves from other social classes (Ogden, 1992). Jewellery can be anything, depending on one's cultural background, and has the power to communicate across and through different generations, as it (jewellery) can be passed on as heirlooms (e.g. Ahde-Deal et al., 2016). Thus, it may communicate beliefs, feelings and the aspirations of people who wore them in the past (Ogden, 1992). This chapter focuses on jewellery in a non-traditional sense, rather than looking at gold necklaces or rings, it centres its attention on *Kente*, a luxury handwoven textile produced in Ghana. *Kente* can be described as a type of jewellery in Ghana, as traditionally, it not only distinguishes social classes but also is often worn as a sign of achievement (e.g. at graduation or formal meetings) and to adorn oneself during festivities (e.g. weddings) (Badoe & Opoku-Asare, 2014; Boateng & Narayan, 2017). *Kente* has a long-standing tradition and plays a vital role in the country's history. Although this luxury cloth is now more widely available, due to cheaper wax prints that imitate the traditional colours and patterns, being able to wear a traditional *Kente* strip produced in Bonwire or the Volta region holds powerful meanings (Smulders Cohen, 2019). In the past, research surrounding *Kente* has predominantly focused on the weavers or the production process of the cloth (ibid.). What remains under-researched is how this luxury item, the *Kente* cloth, can communicate sustainability to conscious consumers, thereby shying away from purchasing cheaper wax prints and investing in the original, handcrafted items.

As highlighted, globalisation implies that textiles can be easily imitated, yet tradition and history, cannot. This is due to the wealth of knowledge that is transferred from generation to generation concerning the culture and practices within the community (Paris, 2020). For these *Kente* communities to survive, it is vital to communicate the meaning to potential consumers that have an interest in both heritage and sustainability.

This chapter explores how far *Kente*, as luxury jewellery, can be used as a communication tool to portray craftsmanship and sustainability along the supply chain, by posing the following research questions:

RQ1: What does sustainability mean within the *Kente* production process?
RQ2: How does *Kente* visually communicate 'sustainability' to the conscious luxury consumer?

This chapter is based on a case study approach—information will be drawn from secondary data to provide an in-depth insight into the traditions and the production processes. It has to be highlighted that the majority of available data stems from the community of Bonwire. Even though other regions are introduced at the beginning of the chapter, the analysis will predominantly focus on Bonwire.

7.2 Background: Bonwire, *Kente*, and *Kente* Production

This chapter investigates the community of Bonwire, located in Western Ghana, as it is the birthplace of *Kente* production, with the community dating back over 300 years (Lartey, 2014; Asmah et al., 2015; Smulders Cohen, 2019). The community itself consists of over 800 houses and is home to approximately 2000 weavers, all of whom are involved in the *Kente* creation process. The weaving profession is generally male dominated, whilst women are more involved in early stages of the production process, such as preparing and dying the yarns in order to be woven into the *Kente* strips, as well as in the later stages of the actual selling of the finished products on the market (Boateng & Narayan, 2017; Smulders Cohen, 2019). Knowledge of the weaving process is passed down from father to son, whilst the spinning, dyeing, and selling process has been passed down from mother to daughter (Boateng & Narayan, 2017).

The focus of this chapter is on *Kente* textiles, which have been declared by the first president of Ghana, Dr Kwame Nkrumah, as the identity for Ghanaians, due to the rich cultural values woven into the individual strips (Fening, 2006). To explain, *Kente* is *the* heritage textile of Ghana and produced either in Western or in Southeast Ghana (Asmah et al., 2015; Smulders Cohen, 2019). The values, beliefs, and traditions are woven into the textiles, providing hidden meanings and story-telling connotations, thus preserving culture through symbols and colours (Badoe & Opoku-Asare, 2014; Kwakye-Opong, 2014; Boateng & Narayan, 2017). Each *Kente* strip has a meaning and philosophy embedded in the patterns and designs (ibid.) that are carefully placed to tell a story and portray standing in society or mark a specific occasion (Boateng, 2011; Boateng & Narayan, 2017). For example, the 'Mako Maso Adeae' pattern holds the same meaning as giving someone a heart necklace or ring, as literally translated it means 'my heart's desire', and proclaims love (Kitenge, 2017). Whilst the *Obaakofo Mmu Man* design is predestined for leaders, its meaning 'one person does not rule a nation' is a reminder that community spirit and working together for the greater good is a key part of the traditions and values of the Ghanaian society (Ross, 1998).

The traditional weaving process of *Kente* strips remains a highly complex process, which involves males using hand-and-foot looms to create carefully designed pieces of woven cloth (Badoe & Opoku-Asare, 2014; Smulders Cohen, 2019). Everyone in the community is involved in the production process, thus, it is a community act that showcases belonging and being part of shaping a culture rich identity. It may not be surprising that the *Kente* textile bears a lot of meaning and is of great value in Ghana, thereby providing Ghanaians with a shared identity (Boateng, 2011; Asmah et al., 2015). Due to the meanings that are attached to colours and symbols woven into the *Kente* cloth, it can act as a form of jewellery, thereby expressing not only feelings (e.g. 'Mako Maso Adeae' pattern), but also political standing (e.g. 'Obaakofo Mmu Man') and social class (e.g. 'Emaa da'—a cloth worn by royalty and those that have a high position within society) (Kitenge, 2017; AdinkraBrand, 2020).

This research examines the stages within the supply chain process of the production of *Kente*, from raw materials to finished products, and identifies what sustainability means. Whether it is solely related to the raw materials used and the finishing processes or if there are other aspects involved that could be classified as 'sustainable'. It further explores how 'sustainability' can be communicated to conscious luxury consumers that have been identified as *Old Money* and *New Ascetics* (Roberts, 2020).

7.3 Sustainability, Supply Chain Management, and Visual Identity

7.3.1 Sustainability

Sustainability has stirred debate since the 1970s, thereby focusing not only on conscious consumption but also on more 'sustainable' business practices, such as changing to more environmentally friendly materials, making supply chains lean, or enforcing tighter social regulations that protect workers (Seuring & Müller, 2008; Henninger et al., 2016; Davies et al., 2020). Sustainability, and more specifically sustainable development, is defined as "meeting needs of the present without compromising the ability for future generations to meet their needs" (UN, 2011). This definition explains that resources should not be overused by the current generation, to the extent that the future generations will have less resources available to meet their needs. Within a community (e.g. Bonwire), there is an aspect of being able to maintain life through the resources we use, in order for future generations to carry on the community practices and traditions (Paris, 2020). It is only through this current generation's maintenance that the future generations may be able to see the importance of their culture and traditions and apply the same wealth of knowledge passed down to them (ibid.). Although this definition has been criticised (e.g. Diesendorf, 2000), it provides a starting point to investigate what needs to be preserved, in the case of this chapter, from the *Kente* weaving process, in order to provide a livelihood for future generations of Bonwire and more specifically the *Kente* weavers. Elkington (1998) divides sustainability into three distinctive pillars: environmental, social, and economic, which are often referred to as the Triple Bottom Line (TBL). These three pillars need to all work together and be in harmony to achieve sustainable development (Kuhlman & Farrington, 2010).

Environmental sustainability implies that a system, here Bonwire's supply chain, utilises raw materials that are less harmful to the natural environment. Thus, it seeks to avoid depletion of non-renewable resources, and overexploitation of renewable resources (Harris, 2003; Kuhlman & Farrington, 2010). Linking this to the *Kente* production supply chain, in order to be more environmentally sustainable, not only raw materials and production processes should be geared towards minimising their impact on the natural environment but also processes should be safe for individuals (e.g. weavers, dyers), by using products that are not harmful (Linton et al., 2007).

Social sustainability focuses predominantly, here, on Bonwire's inhabitants, is concerned with investments in facilities in the community, and makes the place more attractive for people to live (Harris, 2003). Within the supply chain, social practices can further link to labour laws and introduce social standards (Koberg & Longoni, 2019). Here this pillar focuses on how money gained from *Kente* sales is re-invested into the community, thereby also focusing on the needs of future generations.

The economic pillar is concerned with operating a financially viable business that ensures competitiveness in a volatile market environment, (ideally) without harming the natural or social systems (He et al., 2019). In Bonwire, the economic aspect ties in with the fact that the *Kente* production processes have been in operation for over 300 years and are their main income source. It is vital that this source remains viable in the future, to ensure that the inhabitants' livelihoods are not threatened.

In summary, sustainability is the ability to maintain an entity, outcome, or process efficiently over a period of time and this should guide the conduct of individuals in an ethical manner (Gomis et al., 2011; Mensah & Casadevall, 2019). Gomis et al. (2011) consider sustainability to be synonymous with sustainable development. This research looks at sustainability from a 'preservation' angle, which implies that sustainability not simply is seen as something that needs to be maintained but also looks at the past to see what should be preserved. This links to Gibson (2001), who has suggested including culture into the sustainability framework, as cultures differ globally. In this chapter, it is explored, whether preservation goes beyond culture, which can change over time, and suggests that instead heritage should be included, as heritage is based on traditions, values, and beliefs that do not change.

7.3.2 Sustainability and Supply Chain Management

A supply chain and its management are concerned with the entirety of the production process, from gaining raw materials, to producing the product, and finally selling it to the end-consumer (Lambert et al., 2006). Thus, it is "concerned with planning, coordinating and controlling material, parts and finished goods from suppliers to the consumer" (Stevens, 1989, p. 3). Lee (2004) suggests that there are three different qualities of top-performing supply chains: they are **agile**, which means they are rapid in reacting to changing demand, they are able to **adapt** when changes are made in the market structures or when strategies evolve, and they **align** the interest of companies in their supply network in order for the performances in the supply chain to be improved. Thus, the supply chain not only considers the manufacturing of products but also must be able to effectively move them from one point to another (Seuring & Müller, 2008). The overarching goal of supply chain management is to carefully review and potentially change, here, weaving and *Kente* production processes in order to ensure long-term competitiveness (Henninger et al., 2015).

With globalisation, sustainability has increasingly gained a dominant position, also within the jewellery sector (Carrigan et al., 2016). Moreover, sustainable supply chain management (SSCM), defined as "the management of material, information and capital flows (…) while taking goals from all three dimensions of sustainable development, i.e. economic, environmental and social, into account" (Seuring & Müller, 2008, p. 1700), and processes gain momentum (Henninger et al., 2015; Nayak et al., 2020). Gopal and Thakkar (2016) suggest that SSCM aids in the reduction of environmental waste and also considers the social and economic aspects of sustainability. Academics have frequently referred to sustainable supply chain (e.g. Linton et al., 2007; He et al., 2019), and its management (e.g. Koberg & Longoni, 2019; Vijayan & Kamarulzaman,

2020), yet it is unclear what sustainable supply chain processes (SSCP) are. In this chapter, SSCP is the means of being able to balance sustainability factors through activities involved in the flow and transportation of goods from the raw materials to the consumer. This process will involve being able to preserve the environment, as well as consider the people involved within and between these processes and be able to give back to maintain the environment and society.

7.3.3 Sustainability in the *Kente* Production Process

This section focuses on RO1: What does sustainability mean within the *Kente* production process. Looking at the *Kente* production process, and consequently at its supply chain, there are six key stages (Fig. 7.1), which will be carefully reviewed (e.g. Henninger et al., 2015; Smulders Cohen, 2019).

The weaving skill necessary to produce the *Kente* and thus ensure that the supply chain is kept afloat is passed down through generations (Boateng & Narayan, 2017). Interestingly, extant literature discusses the environmental and social aspects of sustainability, but few consider the economic aspect (Gatti & Seele, 2014; Rogge & Reichardt, 2016). Yet, as will be explored, the craftsmanship of the weaver is vital for economic and social sustainability, as without being able to produce the *Kente* products, the community would not be able to exist, as their livelihoods depend on the production process.

Taking the generational aspect into consideration, it becomes clear that the weaving process not only is a tradition but also shapes part of the community's heritage, and thus links to the thought process of 'preservation', which underpins the sustainability framework in this community context. Heritage could be seen to underpin sustainability or to be an element that holds all three pillars together. In the following, the individual stages of the *Kente* supply chain are discussed and explored.

Fig. 7.1 Simplified supply chain process of *Kente* strips (Fening, 2006; Lartey, 2014; Amissah & Afram, 2018; Thirumurugan & Nevetha, 2019)

7.3.3.1 Designing

Design is the first step in the supply chain process. The design aspect refers to the patterns that can be seen on the finished products, as well as the colour selection. As aforementioned, the *Kente* strips not only distinguish individuals of different social classes (Micots, 2020), but are also used for adornment. Different shapes are handwoven into the strips using different colours; each shape and colour carries a different meaning, and they all relate to the life experiences or social standing of Ghanaians (Rovine, 2020). The designs of the *Kente* form a key part in the supply chain process in Bonwire.

There are two aspects to the design process: (1) the design of an individual standalone *Kente* strip, which is used as jewellery and can be draped on the neck or on the arm and worn with a different attire, which makes it stand out; and (2) the design of *Kente* cloth, which is made up of multiple *Kente* strips sewn together to make one big piece of fabric that needs to tell a coherent story (Thirumurugan & Nevetha, 2019) and thus must be carefully joined together to portray the right meaning. This highlights that the weavers must be very skilful to combine the shapes and colours effectively to produce a meaningful product (Fening, 2006). These *Kente* cloths are joined to make up ten yards in length for men to wear draped on their body, and eight yards in length for women to either drape or make into a top or a skirt (Kraamer, 2020).

The art of designing bears a long tradition, in that weavers have been trained by their fathers and grandfathers to visualise the designs in their minds as opposed to sketching them out on paper (Badoe & Opoku-Asare, 2014; Amissah & Afram, 2018). Once the weavers have visualised their designs, a prototype is created, which is presented to the chief for approval, to ensure that they are in line with the traditions and meanings (ibid.). Culturally *Kente* prototypes must be offered to royals first, and only if they decline can be offered to others, which further reiterates the fact that *Kente* strips clearly distinguish social class (Kraamer, 2020).

As aforementioned, cheaper printed versions of the *Kente* design have been produced, as early as the early twentieth century (Halls & Martino, 2018), thereby making the cloth affordable to anyone aspiring to wear similar designs to the handwoven originals. Whilst this highlights that *Kente* has great market potential, it also had consequences for the community, in that cheaper alternatives are more likely to be bought. After gaining political independence from Britain in 1957, *Kente* was made the heritage cloth "as a symbol of Ghana to promote national identity, unity, and pride" (ibid.). Without the weavers passing down their knowledge through generations, without any physical evidence (e.g. sketches), the *Kente* cloth can no longer be produced, and thus key parts of the heritage would be lost. This highlights the importance of heritage to be included as part of 'sustaining' traditions and livelihoods.

7.3.3.2 Sourcing Raw Materials

Kente can contain silk, rayon, or cotton yarns (Lartey & Asma, 2016). Cotton for *Kente* was commonly grown and sourced from Northern Ghana (Lartey & Asma, 2016). With cotton grown in Northern Ghana, raw material transportation was kept at a minimum before reaching its final destination, and thus, the carbon footprint was smaller, as opposed to if cotton were to be purchased from other parts of the world. This links to environmental and economic sustainability: (1) the impact on the natural environment is lessened, by sourcing 'locally' (within the country) (environmental sustainability); (2) the *Kente* weavers were trading kola nuts for cotton, which were needed in other communities to ensure their livelihoods, thereby ensuring that trade stays within the country (economic, social sustainability) (Frimpong & Asinyo, 2013). Yet, with the pressure of companies outside of Ghana producing *Kente*-inspired cloths, the community needed to start 'modernising' by sourcing raw materials outside of their own country, to be able to reduce the overall pricing of the cloth, which implies an increase in carbon emissions.

However, *Kente* weavers have found a way to capitalise on globalisation and reduce the environmental impact, by taking advantage of an ever-growing second-hand market in Africa (James & Kent, 2019). *Kente* weavers are actively sourcing cotton and silk cloth that can be taken apart and reused in their weaving processes (Lartey & Asma, 2016). This encourages a recycling and reuse scheme that fosters sustainability by making use of 'waste' materials rather than harvesting virgin materials.

Although globalisation and mass-produced wax printed *Kente* cloth have threatened Bonwire's *Kente* production, the community has managed to survive, by adapting and capitalising on 'waste', which further reduced costs. A key learning point here is that it is vital to adapt to market challenges and continue to innovate. With sustainability having a centre stage position, the community could further invest in product innovations that may help them to become more circular, by, for example, sourcing raw materials from other waste materials, such as banana peels or mango skins, which have already been trialled and deemed a viable solution with in the industry (Hendriksz, 2017). Whilst it is acknowledged that creating these new innovative fibres might mean a heavy capital investment, collaborations could help to foster economic viability and ensure the community's survival.

7.3.3.3 Spinning the Yarns

The raw materials sourced are spun into yarns on site, which is a predominantly female job (Fening, 2006; Frimpong & Asinyo, 2013). Spinning yearns on site implies increased control, in terms of waste. Women involved in the spinning process ensure that fibre wastage is kept at a minimum (Lartey & Asma, 2016; Thirumurugan & Nevetha, 2019). Similarly, to the design process, the art of spinning yarns has been passed down through generations of Bonwire women (Frimpong & Asinyo, 2013). This links to the aspect of social sustainability, as everyone within Bonwire has a role to play within the production process of the *Kente*. The fact that people have key roles in their community implies that they are also showing pride in their work and look after one another, as the *Kente* production is a community effort. As such, if a community member is unwell and cannot fulfil their task, another person will fill in, if there is a need.

Not all raw materials used in the production process are made from virgin materials, some are also recycled from, for example, second-hand garments; thus, it may not be surprising that the quality (thickness and feel) of the yarns may differ. The quality of the yarns reflects social status, which implies that yarns that are seen to be of lower quality will only be used for designs and patterns that are worn by lower social classes, whilst yarns of good quality may be used to produce *Kente* for royals and high society (Fening, 2006; Micots, 2020).

What becomes apparent from the spinning stage in the *Kente* production process is the fact that sustainability plays a key part, in that all raw materials that are sourced are used and carefully made into yarns, thereby avoiding any fibre loss. Quality issues in the yarns are compensated by producing different products for different audiences (e.g. lower quality, lover price, and lower class).

7.3.3.4 Dyeing

Once the fibres have been spun, the yarns are dyed using plant-based dyes (Lartey, 2014), which is a more environmentally friendly methods than, for example, using industrial dyes that contain chemicals. The dyes, which are often acquired from tree bark, seeds, leaves, and other plants, are sourced in the community's surrounding areas, which make them not only cost efficient (economically viable) but also environmentally friendly (Lartey, 2014; Thirumurugan & Nevetha, 2019).

The dye pits are dug in the ground and carefully maintained to ensure minimum contamination. Once yarns are dyed, the dye pits are carefully covered to ensure that they can be used again when needed (Lartey, 2014). The natural ingredients (e.g. tree bark and seeds) are boiled in water, and the colour pigments are carefully harvested and kept to be used in the dye pits. Depending on the thickness of the yarns, the different plants are used to them, as the colours reflect the social class. Thus, golden colours are used for thick yarns, as these are associated with royal designs and high social standing (Lartey, 2014; Thirumurugan & Nevetha, 2019).

Although dying, similarly to spinning, is a female job, it is strongly linked with the design and weaving process, and thus, strong communication needs to be implemented in order to ensure that cultural meaning is maintained, as certain colours are ascribed to royals and people of wealth (Fening, 2006; Amissah & Afram, 2018).

7.3.3.5 Weaving

The dyed yarns are moved onto the weaving stage. Weaving is a male profession in Bonwire (Ross & Adu-Agyem, 2008), as it is believed that women will be barren if they sit for long hours (Fening, 2006; Micots, 2020). Men are weaving *Kente* using custom-made hand-built looms (Frimpong & Asinyo, 2013; Nunoo et al., 2021), made sustainability from timber. It is important to note that the community of Bonwire is living in harmony with nature, as such the inhabitants are respecting their natural surroundings and seek to have the least impact on it as possible.

The weaving process involves many accessories that are used to achieve the beautiful patterns woven into the strips (Amissah & Afram, 2018). The loom is one of the main accessories to produce the *Kente* strips (Fening, 2006; Badoe & Opoku-Asare, 2014). There is also a heritage aspect here, in that the looms are made specifically out of timber due to its durability and the looms tend to last for years and can be passed down through generations (Fening, 2006).

Today, *Kente* weavers in Bonwire have to ensure that their supply chain is agility as they have to adjust to lesser demand for *Kente* strips and cloth (Lee, 2004). They may have slightly changed some age-old traditions to speed up production processes, but the most important traditions are still kept, like weaving on the loom (Fening, 2006). Although the original dyeing pits still exist and can be used for recycled yarns; to stay competitive, processed dyed yarns are also imported from China and used to weave the *Kente* strips (Boateng, 2011; Boateng & Narayan, 2017). This can reduce lead time of *Kente* cloth for larger orders, as various parts of the supply chain are cut (Sarmiento, 2020). A drawback of globalisation and being able to use these imported yarns is the fact that some of the processes that were traditionally used are threatened to disappear (e.g. sourcing of raw materials, spinning them into yarns and dyeing them).

Although it could be argued that the weaving process could also be under threat due to technologies (e.g. modernised looms), it is unlikely that this tradition will be banished, as the handwoven aspect is what makes the *Kente* a prestigious textile and thus costly. As such, the price implies that these *Kente* strips and cloths are hard to obtain, thereby fostering the meaning and prestige of wearing them (Boateng, 2011).

7.3.3.6 Selling

Selling is the last stage within the *Kente* production. The finished textiles are given to women to sell both within and outside of Bonwire (Amissah & Afram, 2018). The 'made in Bonwire' label that can be given to these *Kente* strips makes them valuable and authentic to those who choose to purchase them, as they are buying into the community effort and the *Kente* heritage. To expand the product range from strips and cloth, *Kente* is now also transformed into bags and ready-made clothing, to attract more consumers (Antwi et al., 2015). Moreover, to ensure that there is no competition among the weavers and the community spirit is fostered, prices for *Kente* products are fixed.

7.4 Visual Identity

Visual identity has been discussed since the 1970s and is associated with visual (tangible) elements of a product, here, *Kente* (e.g. Baker & Balmer, 1997). Visual identity is defined as the outer sign of the inward commitment (Abratt, 1989), in this case, the colours and symbols shown in the *Kente*, which all have different meanings. From Abratt's (1989) definition of visual identity, it shows that the commitment of the internal aspect of Bonwire towards the production of a product is reflected in the product itself. Looking at finished products and the quality of the products, consumers can see how much work and effort has gone into the creation process, which is of interest to Old Money consumers, who seek authenticity and New Ascetics consumers, who have an affinity for artisanry and locally produced products (Roberts, 2020).

Past research (Bolhuis et al., 2018; Tourky et al., 2020; Foroudi et al., 2020) guides this chapter to gain a better understanding of how visual identity is expressed through the *Kente* strip as a piece of jewellery, thereby attracting a new consumer market that could help the community to overcome challenges of globalisation. In this chapter, visual identity is related to the finished products (e.g. strips, cloths, and accessories) and more specifically what stories these products tell a consumer, in terms of symbolism, and the production process. As aforementioned, *Kente* production is a community effort, with long traditions, thus it could be said that each strip tells a unique story, as it has been produced by an individual of the community, and reflects the experiences of inhabitants through the colours and designs (Ross & Adu-Agyem, 2008).

With the increased threat of mass-produced *Kente*-inspired products circulating on the market, it is vital for Bonwire as a community to still be able to sell their traditional cloths, to ensure they are financially viable, and thus attracting consumer types such as Old Money and New Ascetics consumers can be an opportunity. Some of the most prominent elements of visual identity reflected in *Kente* strips/cloths are the shapes and symbols used in the textiles, as well as the colour scheme, where each colour has a specific meaning. Colour plays an important role in bringing out responses from audiences and highlights the character or the status of the person wearing the *Kente* materials (Fening, 2006; Dor, 2014).

7.4.1 *Kente*'s Visual Identity and How It Communicates Sustainability to Luxury Consumers

This section addresses RO2: how does *Kente* visually communicate 'sustainability' to the conscious luxury consumer.

Luxury customers, when making purchase decisions, usually look for social, religious, economic, or demographic factors within the region to influence their decisions to make a purchase (Obeidat & Young, 2016). Winning luxury consumers depends on a brand's ability to appeal to the local culture and take into consideration the local religion and values of that place, especially in advertising (Obeidat & Young, 2016). In the case of *Kente*, considering the processes involved in the supply chain could make it especially appealing for Old Money and New Ascetics consumers (Roberts, 2020).

All processes described in Sect. 7.2 of the supply chain can also be observed in the *Kente* Weaving Centre, a tourism centre that was built in Bonwire to showcase traditions and heritage and ensure an 'authentic' feel for the products, which may be especially attractive for both new types of luxury consumers, as it is both authentic and locally made (Amissah & Afram, 2018; Roberts, 2020). According to Obeidat and Young (2016), luxury consumers consider cultural heritage in products they want to purchase (e.g. New Ascetics). This would put *Kente* products on the map, as they are full of cultural heritage and embedded with different philosophies and stories about the experiences in the life of those that wear them. In the case of *Kente* strips, the fact that they are handwoven will make the product come across as being a sustainable product. There is no evidence of machinery used to automatically weave the strips and everything is manually made by the inhabitants of the community, and thus authentic, which might speak to Old Money consumers.

As highlighted, culture, values, and beliefs have been woven into the *Kente* designs and different colours and shapes hold different meanings (Dor, 2014). This is the same in the case of jewellery, in other countries, and as such, *Kente* is usually used as an heirloom (Brown, 2020). In the case of *Kente* strip as a form of jewellery, the visual identity may play a role in the supply chain process. This means that there is a desired outcome for the product and, as such, the production process will contribute to this visual identity.

7.4.2 Visual Identity, Sustainability, and Supply Chain Processes

An aspect that has previously not been researched is the link between visual identity and SSCP. Yet, in the case of *Kente*, a link can be observed as the first stage of the supply chain (design) visualises the history, traditions, and values of the community, which is carefully portrayed through symbols and colours (Dor, 2014). As was illustrated in Sect. 7.2, sustainability is reflected throughout the traditional supply chain process of the *Kente* production, which is underpinned by long-standing traditions, and thus heritage. Raw materials are carefully sourced, dyed, and processed to reduce waste. Although some of the processes are changing to make it more economically viable, being able to purchase a traditional stripe holds value to consumers, as it can be seen as an investment piece, an heirloom that can be handed down through generations. The symbols and colours portray a hidden meaning that is only known by individuals accustomed with the Ghanaian culture, and as such making it an authentic piece that portrays local artisanry, key aspects that are sought after by Old Money and New Ascetics consumers (Roberts, 2020).

7.5 Conclusion

In summary, it can be said that the community of Bonwire is currently at a turning point. It has managed to survive the industrial revolution and other technological challenges, thereby remaining true to its traditions and heritage. The latter is especially important, as it highlights sustainability can mean more than simply being environmentally friendly, economically viable, or socially responsible, it implies in order to be 'sustainable' traditions need to be kept alive, as it is these traditions that hold together the community and ensure that knowledge is passed down through generations, thereby allowing consumers to indulge in products that are meaningful and can carry hidden symbols.

Whilst mass-produced *Kente*-inspired prints can help to raise awareness of the *Kente* designs, showcasing the supply chain, highlighting the sustainable practices, and artisanry of the country are meaningful to the new types of luxury consumers that are on a quest to not simply acquire more luxury, but something that is authentic and meaningful. As such, it could be argued that incorporating heritage into the sustainability framework provides communities, such as Bonwire, with a competitive edge that might make them survive for another 300 years to come.

Luxury textile companies, in order to remain meaningful or survive, might consider transparency of the supply chain. This allows consumers to see how the knowledge of manufacturing textiles is passed down to employees. Luxury textile companies should also consider its history and how it started, adopting some old techniques which made their establishment meaningful and useful to the masses. Consumers may look for a story to be a part of, just as in the case of Bonwire; wearing a piece of woven textile with a particular design says much about the person wearing it.

References

Abratt, R. (1989). A new approach to the corporate image management process. *Journal of Marketing Management, 5*(1), 63–76.
AdinkraBrand. (2020). *Kente: Patterns, symbolism and meaning*. AdinkraBrand (online). Retrieved June 07, 2020, from https://www.adinkrabrand.com/blog/kente-cloth-everything-you-need-to-know-about-africas-most-iconic-fabric/
Ahde-Deal, P., Paavilainen, H., & Koskinen, I. (2016). It's from my grandma. How jewellery becomes singular. *The Design Journal, 20*(1), 29–43.
Amissah, E. K., & Afram, A. P. (2018). A comparative study of Bonwire Kente and Daboya Benchibi. *Trends in Textile and Fashion Design, 1*(5), 96–108.
Antwi, A. P., Bin, C., Tetteh, A., & Adashie, M. (2015). Consumers preference and further uses of Kente cloth. *Art and Design Studies, 37*, 36–43.
Asmah, A. E., Gyasi, I., & Daitey, S. T. (2015). Kente weaving and tourism in a cluster of Kente towns in Ashanti. *International Journal of Innovative Research and Development, 4*(11), 113–120.
Athwal, N., Wells, V., Carrigan, M., & Henninger, C. E. (2019). Sustainable luxury marketing: A synthesis and research agenda. *International Journal of Management Review, 21*(4), 405–426.
Badoe, W., & Opoku-Asare, N. (2014). Structural patterns in Asante *Kente*. *Journal of Education and Practice, 5*(25), 52–64.
Baker, M., & Balmer, J. M. T. (1997). Visual identity: Trappings or substance? *European Journal of Marketing, 31*(5/6), 366–382.
Belk, R. (1988). Possessions and the extended self. *Journal of Consumer Research, 15*(2), 139–168.
Boateng, B. (2011). *The copyright thing doesn't work here*. University of Minnesota Press.
Boateng, H., & Narayan, B. (2017). Social capital and knowledge transmission in the traditional Kente textile industry of Ghana. *Information Research, 22*(4), 1–19.

Bolhuis, W., de Jong, M. D. T., & van den Bosch, A. L. M. (2018). Corporate rebranding: Effects of corporate visual identity changes on employees and consumers. *Journal of Marketing Communications, 24*(1), 3–16.
Brown, A. (2020). Four ethical luxury jewellers to put on your radar. *Financial Review* (online). Retrieved June 05, 2020, from afr.com/life-and-luxury/fashion-and-style/four-ethical-luxury-jewellers-to-put-on-your-radar-20200421-p54lss
Carrigan, M., McEachern, M., Moraes, C., & Bosangit, C. (2016). The fine Jewellery industry: Corporate responsibility challenges and institutional forces facing SMEs. *Journal of Business Ethics, 143*, 681–699.
Davies, I., Oates, C. J., Tynan, C., Carrigan, M., Casey, K., Heath, T., Henninger, C. E., Lichrou, M., McDonagh, P., McDonald, S., McKechnie, S., McLeay, F., O'Malley, L., & Wells, V. (2020). Seeking sustainable futures in marketing and consumer research. *European Journal of Marketing* (ahead of print).
Diesendorf, M. (2000). Sustainability and sustainable development. In D. Dunphy, J. Benveniste, A. Griffiths, & P. Sutton (Eds.), *Sustainability: The corporate challenge of the 21st century* (pp. 19–37). Allen & Unwin.
Dor, G. W. K. (2014). Ephraim Amu's "Bonwere Kenteŋwene": A celebration of Ghanaian traditional knowledge, wisdom and artistry. *African Music: Journal of the International Library of African Music, 9*, 7–35.
Elkington, J. (1998). Partnerships from cannibals with forks: The triple bottom line of 21st century business. *Environmental Quality Management, 8*(1).
Fening, K. O. (2006). History of Kente Cloth and its value addition through design integration with African wild silk for export market in Ghana. In *Paper presented at trainers course and 4th international workshop on the conservation and utilisation of commercial insects, Duduville, Nairobi*, Vol. 1, pp. 62–66.
Foroudi, M. M., Balmer, J. M. T., Chen, W., Foroudi, P., & Patsala, P. (2020). Explicating place identity attitudes, place architecture attitudes and identification triad theory. *Journal of Business Research, 109*, 321–336.
Frimpong, C., & Asinyo, B. K. (2013). A comparative study of history, equipment, materials, techniques and marketing approach in the traditional weaving in Ghana. *Art and Design Studies, 7*, 1–8.
Gatti, L., & Seele, P. (2014). Evidence for the prevalence of the sustainability concept in European corporate sustainability reporting. *Sustainability Science, 9*, 89–102.
Gibson, R. B. (2001). *Specification of sustainability-based environmental assessment decision criteria and implications for determining "significance" in environ-*

mental assessment. Canadian Environmental Assessment Agency Research and Development Programme, Ottawa, Canada.

Gomis, A. J. B., Parra, M. G., Hoffman, W. M., & McNulty, R. E. (2011). Rethinking the concept of sustainability. *Business Society Review, 116*(2), 171–191.

Gopal, P. R. C., & Thakkar, J. (2016). Sustainable supply chain practices: An empirical investigation on Indian automobile industry. *Production Planning and Control, 27*(1), 49–64.

Halls, J., & Martino, A. (2018). Cloth, copyright, and cultural exchange: Textile designs for exports to Africa at the National Archives of the UK. *Journal of Design History, 31*(3), 236–254.

Harris, J. M. (2003). *Sustainability and sustainable development*. International Society for Ecological Economics (online). Retrieved May 05, 2020, from http://isecoeco.org/pdf/susdev.pdf

He, Q., Gallear, D., Ghobadian, A., & Ramanathan, R. (2019). Managing knowledge in supply chains: A catalyst to triple bottom line sustainability. *Production Planning and Control, 30*(5/6), 448–463.

Hendriksz, V. (2017). *Sustainable textile innovations: Banana fibres*. FashionUnited (online). Retrieved June 25, 2020, from https://fashionunited.co.uk/news/fashion/sustainable-textile-innovations-banana-fibre/2017082825623#:~:text=Fabrics%20made%20from%20banana%20fibres,alternative%20to%20coton%20and%20silk

Henninger, C. E., Alevizou, P. J., Oates, C. J., & Cheng, R. (2015). Sustainable supply chain management in the slow-fashion industry. In T. M. Choi & T. C. E. Cheng (Eds.), *Sustainable fashion supply chain management: From sourcing to retailing* (pp. 83–100). Springer.

Henninger, C. E., Alevizou, P. J., & Oates, C. J. (2016). What is sustainable fashion? *Journal of Fashion Marketing & Management, 20*(4), 400–416.

James, A. S. J., & Kent, A. (2019). Clothing sustainability and upcycling in Ghana. *The Journal of Design, Creative Process and the Fashion Industry, 11*(3), 375–396.

Kitenge. (2017). *Everything you need to know about Kente*. Kitenge (online). Retrieved May 13, 2020, from https://kitengestore.com/everything-need-know-kente/

Koberg, E., & Longoni, A. (2019). A systematic review of sustainable supply chain management in global supply chains. *Journal of Cleaner Production, 207*, 1084–1098.

Kraamer, M. (2020). A cloth to wear: Value embodied in Ghanaian textiles. In R. Granger (Ed.), *Value construction in the creative economy* (Palgrave Studies in business, Arts and Humanities). Palgrave Macmillan.

Kuhlman, T., & Farrington, J. (2010). What is sustainability? *Sustainability, 2*(11), 3436–3448.

Kwakye-Opong, R. (2014). Beyond ethnic traditions: Philosophies and sociocultural relevance of the Ashanti and Ewe Kente cloths. *Research on Humanities and Social Sciences, 4*(26), 150–165.

Lambert, D. M., Croxton, K. L., Garcìa-Dastugue, S. J., Knemeyer, M., & Rogers, D. S. (2006). *Supply chain management processes, partnerships, performance* (2nd ed.). Hartley Press.

Lartey, R. L. (2014). *Integrated cultural weaves (Fugu, Kente and Kete) woven with organic dyed yarns*. Master's Thesis, Kwame Nkrumah University of Science and Technology, Ghana.

Lartey, R. L., & Asma, A. E. (2016). Organic dyed yarns for Kente weaving. *Africa Development and Resources Research Institute Journal, 11*(3), 18–42.

Lee, H. L. (2004) The triple-a supply chain. *Harvard Business Review* (online). Retrieved June 25, 2020, from https://hbr.org/2004/10/the-triple-a-supply-chain

Linton, J. D., Klassen, R., & Jayaraman, V. (2007). Sustainable supply chains: An introduction. *Journal of Operations Management, 25*(6), 1075–1082.

Mensah, J., & Casadevall, S. R. (2019). Sustainable developments: Meaning, history, principles, pillars and implications for human action: Literature review. *Cogent Social Sciences, 5*, 1–21.

Micots, C. (2020). Kente cloth (Asante and Ewe peoples). *Khan Academy*. Retrieved June 20, 2020, from https://www.khanacademy.org/humanities/art-africa/ghana/a/kente-cloth

Nayak, R., Akbari, M., & Far, S. M. (2020). Recent sustainable trends in Vietnam's fashion supply chain. *Journal of Cleaner Production, 225*, 291–303.

Nunoo, S., Parker-Strak, R., Blazquez Cano, M., & Henninger, C. E. (2021). My loom and me: The role of the handloom in a Weaver's identity creation. In M. A. Gardetti & S. S. Muthu (Eds.), *Handloom sustainability and culture*. Springer.

Obeidat, M., & Young, W. D. (2016). Consumer purchasing decision for fashion luxury brands in Dubai: A case of Armani. *International Journal of Management, Accounting and Economics, 2*(4), 7–23.

Ogden, J. (1992). *Ancient jewellery: Interpreting the past*. The University of California Press.

Paris, D. Y. (2020). *We are the Earth: How cooperation with indigenous communities can change the World.* Eco-Age (online). Retrieved June 26, 2020, from https://eco-age.com/news/how-cooperation-with-indigenous-communities-can-change-the-world

Rickenbacher, P. (2020). *Forget retail therapy – This is the age of the conscious consumer.* World Economic Forum (online). Retrieved May 13, 2020, from https://www.weforum.org/agenda/2020/01/conscious-consumption-not-retail-therapy/

Roberts, F. (2020). *Global luxury goods: Today and tomorrow.* Euromonitor International (online). Retrieved May 13, 2020, from http://www.dragonink.co.th/trend2013/Global_Luxury_Good%20Today&Tmr.pdf

Rogge, K. S., & Reichardt, K. (2016). Policy mixes for sustainability transitions: An extended concept and framework for analysis. *Research Policy, 45*(8), 1620–1635.

Ross, D. H. (1998). *Wrapped in pride: Ghanaian Kente and African American identity.* UCLA Fowler Museum of Cultural History.

Ross, M., & Adu-Agyem, J. (2008). The evolving Art of Ashanti Kente weaving in Ghana. *Art Education, 61*(1), 33–38.

Rovine, V. L. (2020). Woven beliefs: Textiles and religious practice in Africa. In V. Narayanan (Ed.), *The Wiley Blackwell companion to religion and materiality.* Wiley.

Ryding, D., Henninger, C. E., & Blazquez Cano, M. (Eds.). (2018). *Vintage luxury fashion: Exploring the rise of secondhand clothing trade.* Palgrave Advances in Luxury Series, Palgrave.

Sarmiento, I. G. (2020). *Kente cloth: From royals to graduation ceremonies…to congress?* NPR (online). Retrieved June 20, 2020, from https://www.npr.org/sections/goatsandsoda/2020/06/11/875054683/kente-cloth-from-royals-to-graduation-ceremonies-to-congress

Seuring, S., & Müller, M. (2008). From a literature review to a conceptual framework for sustainable supply chain management. *Journal of Cleaner Production, 16,* 1699–1710.

Smulders Cohen, J. (2019). The *Kente* weavers of Ghana. *Text, 17*(2), 149–157.

Stevens, G. C. (1989). Integrating the supply chain. *International Journal of Physical Distribution & Materials Management, 19*(8), 3–8.

Thirumurugan, V., & Nevetha, R. P. (2019). A review article on "Kente cloth in home furnishings" – overview. *Journal of Textile Engineering and Fashion Technology, 5*(6), 306–308.

Tourky, M., Foroudi, P., Gupta, S., & Shaalan, A. (2020). Conceptualizing corporate identity in a dynamic environment. *Qualitative Market Research*.

UN (United Nations). (2011) *Report of the world commission on environment and development: Our common future.* UN Documents (online). Retrieved June 09, 2020, from http://www.un-documents.net/wced-ocf.htm

Vijayan, G., & Kamarulzaman, N. H. (2020). An introduction to sustainable supply chain management and business implications. In Information Resource Management Association (Ed.), *Sustainable business: Concepts, methodologies, tools and applications.* IGI Global.

8

Canadian Ethical Diamonds and Identity Obsession: How Consumers of Ethical Jewelry in Italy Understand Traceability

Linda Armano and Annamma Joy

8.1 Introduction

Product traceability allows retailers and consumers to know where a product was produced and to trace its journey through tiers of suppliers until it is a completed product offered for sale. Thus, traceability embodies such concepts as supply chain, ethics, sustainability, transparency, and informed choice (Gurzawska, 2020). In the jewelry industry, the most well-known traceability systems in relation to gold and precious stones

---

L. Armano
Faculty of Management, University of British Columbia Okanagan, Kelowna, BC, Canada

Department of Management, Università Ca' Foscari Venezia, Venice, Italy
e-mail: larmano@ubc.ca

A. Joy (✉)
Faculty of Management, University of British Columbia Okanagan, Kelowna, BC, Canada
e-mail: annamma.joy@ubc.ca

involve specific certification programs (e.g., Fairmined gold) and traceability schemes substantiated by laser engravings of logos and alphanumeric codes on stones (e.g., Canadian ethical diamonds, or CEDs). The concept of traceability is commonly defined as "the ability to identify and trace the history, distribution, location and application of products, parts, materials and services" (Garcia Torres et al., 2019, p. 85). Such traceability is often seen as an indicator of quality (Bloemer et al., 2009) and as a feature able to influence consumption choices (Roth & Diamantopoulos, 2009). The issue of blood diamonds (also termed *conflict diamonds*) has underscored the significance of traceability. Mined in civil war zones, often in Angola, Congo, Ivory Coast, and Sierra Leone, among others, such diamonds have been used to finance war-related activities, as recounted in numerous reports (D'Angelo, 2019), and featured in the plots of a multitude of television shows and movies (e.g., 2002's *Die Another Day* and 2006's *Blood Diamond*). Tiffany's has made its commitment to using only responsibly sourced diamonds in its jewelry collections a hallmark of its marketing campaigns and corporate ethos (https://www.tiffany.com/sustainability/).

Within the past two decades, many governments have taken action to curtail the flow of diamonds from conflict zones through special certificates. In 2003, an international initiative, the Kimberley Process Certification Scheme (KPCS, known familiarly as KP), was created in 2003, with the mission of ensuring that the sale of rough stones exported by signatory nations did not finance civil conflicts or actions of international terrorism. Thus, KP-certified rough diamonds are still only exported and imported, within sealed packages, between countries that have joined the KP (Tripathi, 2010). However, this certification can be susceptible to fraud, as it does not provide a totally secure solution to stopping the flow of illegally sold diamonds; among other issues, the KP certification tracks rough diamonds only through the polishing process, rather than through to the final product being available for sale (McManus et al., 2020).

Canada, home to an active and growing mining industry in the Northwest Territories (NWT), has been involved with the KP from its inception, with the distinction of exporting diamonds mined in absolute legality; because Canada is free of civil wars, there are no links between mining and conflicts (McManus et al., 2020). In addition to their KP certification, Canadian diamonds carry a certification signed by the Government of the Northwest Territories (GNWT) that labels stones mined in NWTs as ethical diamonds, allowing consumers to know, unequivocally, not only the country of origin (CoO) but also the Mine of Origin (MoO) of the diamonds (Tripathi, 2010). The Canadian diamond industry is thus a standard-bearer for ethical consumption (Gurzawska, 2020) which, in turn, supports other values such as respect for the rights of the workers (e.g., work safety protocols, fair wages, and the like) involved along the diamond supply chain from the mine to the consumer (Tripathi, 2010); and the safeguarding of the environment via monitoring by mining multinationals in the NWT of mine-related pollution.

While most studies on the topic of traceability concern analyses of the discursive strategies of advertising campaigns on ethical jewelry, the present research, based on an ethnographic survey of Italian ethical jewelers and representative customers, focuses on how the topic of traceability is communicated and negotiated within two particular Italian ethical jewelry stores, Belloni Jewelers in Milan and Simone Righi Jewelers in Bologna.

The current study is part of a broader research project, in progress as of this writing, whose purpose is to explore, within a global context and using a multi-sited ethnographic survey, the cultural interpretations that different subjects (miners, the staff of multinational mining companies, members of Indigenous communities, consumers, and jewelers) give to the concept of CEDs extracted from the mines of Ekati and Diavik in the NWT.

Belloni Jewelers, a pioneer of ethical jewelry retailing in Italy, was the first store in the country to sell ethical diamonds mined in Canada, starting in 2005. In 2010, the store became a wholesaler of these diamonds in Italy; the Simone Righi Simone jewelry store in Bologna was the first of Belloni's customers to order them. While other Italian jewelers (in Rome and Florence) now purchase CEDs from Belloni Jewelers, we chose to focus on Belloni and Simone Righi because of their respective emphasis on traceability in diamond mining and manufacture. Our preliminary study aimed to analyze how information about the traceability of CEDs successfully mitigates consumer concerns about the ethics of jewelry supply chains, and to explore the role of what Remotti (2010) terms identity obsession (referred to hereinafter as IO) in the purchase decisions these consumers reach related to CEDs.

8.2 Literature Review

The issue of diamond traceability is frequently perceived as related to information governance (Bailey et al., 2016), in which information about a product, very often established by governments and corporations, undergoes a series of transnational transfers within value chains (Coff et al., 2008). Extant literature discusses such concepts as traceability for sustainability (TFS) (Garcia Torres et al., 2019) and ethical traceability (ET) (Coff et al., 2008), particularly in the context of luxury goods, to explain the need for consumers to be educated not only about the material aspects of a given product but also about its impact sociologically (e.g., workers' rights, community needs, and animal welfare) and environmentally (Bradu et al., 2013). A focus on consumers' perception of luxury product traceability (Romani et al., 2013) notwithstanding, much remains unexplored; traceability is a relatively new concern. Tiffany, for example, only garnered the ability to trace all of the rough diamonds used in its product lines to known mines or reputable suppliers working with known mines in 2019 (https://www.international.tiffany.com/sustainability/product/). Such actions on the part of a celebrated luxury brand reflect evolving perceptions of the role of consumption, with the purchase of luxury items no longer experienced as solely an expression of desire to possess beauty, and an appreciation of that beauty; rather, the act of consumption itself now sites individuals within a nexus of globally relevant social and environmental obligations. (https://www.tiffany.com/engagement/diamond-provenance/). Marin et al. (2009) argue that consumer purchase decisions of luxury goods are influenced by guarantees of traceability, leading to brand loyalty and willingness to serve as informal brand ambassadors. Du et al. (2010) suggest that traceability, as a primary component of corporate social responsibility (CSR), is essential in creating a dialogue between companies and consumers, while Bhattacharya et al. (2009) see a pressing need for a more precise understanding on the part of stakeholders and consumers of the underlying processes driving the narrative of a product's traceability and subsequent consumer purchasing behavior. Such urgency is particularly relevant in relation to luxury goods and to understanding the underlying processes driving the narrative of traceability.

Any marketing narrative, whether in the context of luxury goods or other products, is designed to elicit an emotional reaction in consumers; in luxury goods, however, the elicitation is intensified, given that consumers typically purchase luxury items out of desire rather than need. However, when a product that is inherently unnecessary can be positioned as a force for good, its appeal can be appreciated on multiple levels, transforming an action of indulging the self into one of enhancing the self via the demonstration of morality, thereby conferring virtue on the individual who has chosen to take this action. A luxury brand can thus serve as an instigator for positive moral decisions (e.g., purchasing jewelry featuring a CED) that highlights the ideals of the brand's customers (Escalas, 2004).

In our study, by focusing on the storytelling of Canadian diamond traceability as told by jewelers to Italian consumers, we investigate a specific narrative built around CEDs that primarily leverages customer concerns about the violation of the safety and human rights of workers on the diamond supply chain. Previous studies suggest that the narrative of the traceability of CEDs focuses on the protection of workers hired by multinational mining companies in NWTs and by companies offering collateral services to mining companies (Armano & Joy, 2021). Unlike CED narratives, those relating to the traceability of other precious metals such as gold (e.g., Fairmined gold) emphasize pollution and community harm resulting from gold mining rather than workers' rights (ibid.).

8.2.1 Obsessive Identity

By using the concept of IO (Remotti, 2010) to understand how the issue of traceability is explained by jewelers and perceived by consumers, we see how traceability is the key element in determining jewelry customers' decisions to purchase CEDs rather than non-certified diamonds. Remotti states that individuals depend on the concept of identity in all areas of their lives (Bauman 2007). Remotti further argues that this obsession with identity is a specific component of contemporary times, and within the social sciences, which, until the 1960s, were concerned with such concepts as alienation, dialectics, and structure rather than identity. The concept of identity can be expressed in Remotti's (2010) formula A = A: "If I say that this clock is this clock I express the most indisputable truth of this world, I express an absolute certainty. The principle of identity is accompanied by the principle of non-contradiction, whereby A is not only equal to A (A = A), but is different from anything that is not A (A ≠ not A)" (p. 3). Tracing the history of European ontological thought since the seventeenth century, Remotti further argues that the concept of identity has gradually been assumed as a psychological tool necessary for reassurance and certainty. In line with several philosophers of the 1700s (i.e., David Hume), the author further argues that in order to identify something, two essential elements are necessary: memory and imagination. In line with Remotti, Park et al. (1994) and Brucks (1985) suggest that for consumers to truly understand the facts they learn about product traceability requires a competence comprising subjective and objective knowledge, which will have already been consolidated and will therefore allow consumers to recognize and store this information in memory. However, because memory by its nature has gaps (Remotti, 2010), it is insufficient to reconstruct, on its own, the history of CEDs. Therefore, imagination fills the gap, ensuring that the path taken by the diamond from the MoO to the jeweler's display case is clearly understood. Customers simply trust the information given them by the jewelers, with their imaginations activated by the storytelling about traceability fleshing out the facts with which they have been presented. It must be emphasized, however, that the customers we interviewed did not fully accept the traceability narrative as an objective certainty, but rather as a highly probable assumption. Nevertheless, the accuracy and authority of the information conveyed by jewelers through CEDs' traceability storytelling ensures that this narrative is perceived as a sufficiently reliable source to guide consumers' purchase choices (Chen & Huang, 2013). Concurrently, storytelling about ethical jewelry allows consumers to attribute an inherently higher quality to CEDs that differentiates them from non-certified diamonds. The materially higher quality of CEDs may in fact be visually distinct from non-CEDs, since a diamond's characteristic flash and prismatic color will likely be brighter and have more depth when the cutting, polishing, and setting of a diamond are done by skilled artisans, who are typically more likely to experience good working conditions than their less skilled counterparts. However, such differentiation may not be apparent to an untrained eye; an experienced jeweler will see what an inexperienced customer will not.

8.3 Methodology

To understand how storytelling narratives about the traceability of CEDs drive customers' purchase choices and how the application of the IO concept allows us to understand such behaviors, our ethnographic research, following an interpretive approach (Joy et al., 2014), was conducted from 2020 through early in 2021, in the ethical jewelry stores Belloni in Milan and Simone Righi in Bologna. Fourteen semi-structured interviews were conducted with both jewelers (Francesco Belloni and Simone Righi) and 12 of their customers, predominantly male, who purchased jewelry featuring CEDs. The owner of Belloni jewelers helped us in contacting some of his customers who were willing to be interviewed. Their ages ranged from 30 and over; all were Italians by birth and currently resident in northern and central Italy, except one interviewed who was resident near Naples. All interviews lasted about 1 h and were conducted directly in the stores. We also visited jewelry stores in Milan and Bologna where jewelers showed us some articles of ethical jewelry.

8.4 Findings

8.4.1 Analysis of Ethnographic Data

The storytelling about CED traceability incorporates and articulates the following: the meaning of the MoO, which offers consumers assurance both of a given provenance, and that environmental norms were not violated; the knowledge of the material characteristic of given stones, from the rough diamond to the final product; the guarantee that mining in Canada has no connection to funding conflict, and the certainty, according to those we interviewed, that regulations for worker safety were applied throughout the diamond supply chain. In addition, traceability allows diamond mining to be located within an economically developed and industrialized mining country such as Canada, rather than in economically poor countries in Africa such as Botswana and Namibia, and gives interlocutors and consumers the perception of higher value, not so much material as intangible, of Canadian diamonds compared to other diamonds (Brun et al., 2008).

Because CEDs are not a common purchase in Italy, broadening the target market will require a narrative that helps consumers—those already sensitive to sustainability and ethics and those who are not—to fully understand the difference between CEDs and diamonds mined in conflict areas. The fewer consumers know about the pitfalls of the diamond trade, the more they need to be informed of the benefits of ethical diamonds, through a narrative built ad hoc whose main theme is the traceability of the stone.

8.4.2 Traceability and Communication Strategies

During our interview with the Milanese jeweler Francesco Belloni, the challenge of positioning CEDs within the jewelry industry emerged:

"In Italian jewelry fairs, it is impossible to talk about ethical diamonds. There is great difficulty in spreading this alternative to ordinary diamonds. There is a sort of code of silence on this issue" (Francesco

Belloni). The questions that arose in our minds were as follows: why is there a code of silence? Why are other jewelers in Italy resistant to CEDs? We are aware that the introduction of new products is, in general, a complicated challenge; indeed, it follows complex processes of interaction among people, knowledge, policy, and market. Notably, as regards a niche product, like CED, it may stay at the fringes of the luxury market and have difficulties being culturally assimilated. In Italy, we noted that the jewelry market is held by a few big business groups who can control the entire market. It becomes clear that various different actors (jewelry companies, jewelers, as well as consumers) directly or indirectly hinder the successful introduction of new varieties of diamonds.

To introduce CEDs to potential customers, Belloni and Simone Righi deploy a communication strategy that emphasizes the assumption of respectful behaviors in all actions, modeling such behavior (in the case of Belloni), through charities (e.g., donations to associations that assist cancer patients), funded by ethical jewelry purchases in their stores (Armano & Joy, 2021). In fact, as jeweler Simone Righi of Bologna explained to us, the business of selling ethical jewelry relates to his personal values:

> My path as a man has been to be a conscientious objector. That's why I don't want to sell diamonds that finance civil wars. Also, I am a volunteer in my city and help people who are economically needy. I use renewable sources in my company. Insofar as I can, I try to operate as fairly as possible. In the area of jewelry, I, therefore, support those who provide alternative options. The choice to sell ethical diamonds was a natural one for me.

Storytelling that broadcasts and reflects the jewelers' own personal values connects them on a highly personal level to the product they sell. In doing so, the narrative that leverages traceability allows consumers, with the aid of the emotional-narrative transport of the jewelers, to become familiar with the products. Indeed, some authors (Spinelli et al., 2015; Graulau, 2008) note that the communication strategy with which a product's storytelling is constructed is typically focused by 80 percent on emotional content; only the remaining 20 percent attempt to build and expand brand awareness (Danner et al., 2017). For niche products such as ethical jewelry, the narrative theme about traceability, which also evokes MoO (Graulau, 2008), is crucial to emotional reassurance and thereby drives consumers toward the choice of CEDs.

How can the meaning of traceability be conveyed concretely to consumers? In addition to evangelizing on the subject for his customers, Belloni regularly makes appearances at educational facilities to discuss ethical jewelry:

> I have been invited to many design schools in Milan that are famous all over the world… the Milan Polytechnic, Catholic University, and the Brera Academy in Milan. There I explained to the students the possibility of designing their own jewels incorporating the logic of ethics. In this way they can distinguish themselves from others and can also have more income. (Francesco Belloni)

Both jewelers agree that information about CEDs should be available not only in their stores but also at conferences within fair trade fairs. While CED narratives offered in jewelry stores have a ready audience, since store customers have already expressed interest in CEDs, attendees at professional trade conferences may be new to the topic. The strategy of using the conference as a channel for disseminating data, in addition to reaching a wide audience, gives the information formality and thus greater narrative power, as the topic attains cultural as well as commercial ramifications.

While jewelers at professional conferences will certainly be knowledgeable about diamonds, if not about CEDs, customers in jewelry stores may be in the dark about both. As one Belloni Jewelers customer stated:

> I don't understand anything about diamonds. However, I have always been a customer of Belloni Jewelers. Here I came to know that the Canadian ethical diamond is produced following professional ethics throughout the supply chain in which exploitation of all the workers involved is avoided. The finished product costs more because of these safeguards and I trust this information. Nevertheless, I don't know the mines where these diamonds are extracted. (Luca, Milan, 30 years old)

This discourse, schematized in Fig. 8.1, makes it possible to unite the theme of CEDs traceability, ethical consumer trust in the information told by jewelers, and the concept of IO.

The knowledge gap about diamonds and MoO shown in Fig. 8.1 is filled by the imagination of the customer to whom the CEDs story is told, as another Belloni Jewelers customer explained:

> When I went to Belloni Jewelers, the Canadian diamond story was explained to me so all the qualms I had fell away. When I returned home, I… read [more] information on the Internet. Then I returned to the jewelry store to pick up the ring I had ordered and with the jeweler we delved into some aspects of mining. It was this approach that convinced me to purchase a piece of jewelry that had an ethical diamond mounted on it. I have always known that many diamonds come from Africa, where there are inhumane working conditions. For this reason, I have always been very apprehensive about buying diamonds. (Gustavo, 39, Engineer, Monza)

Gustavo's testimony reveals the relationship between the consumer's fear of making a wrong purchase choice and the trust engendered after listening to the storytelling about CEDs and gaining objective knowledge of the product. Many interviewees in our study had read online reports of working conditions in African mines. Taleb (2007) suggests that many people tend to educate themselves on a given subject by approaching it from subject areas with which they are already familiar. Armano and Joy (2021) demonstrate that the Italian ethical consumers they interviewed tended to compare ethical jewelry (unfamiliar) to Made in Italy products (extremely familiar), attaining comfort with the former by linking it to the latter. Our interviewees sought education through research that employed exclusionary thought processes: "I can't imagine Canadian mines, but I can say how I imagine African mines. Then from these I can guess how diamonds are extracted in Canada" (Umberto 50, Architect, Milan). Other Belloni Jewelers customers also gave a summary image of Canadian and African mines reconstructed through fragments of information retrieved from various sources:

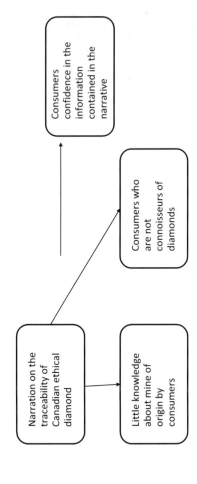

Fig. 8.1 Conveyance of the concept of ethical diamond traceability to Italian ethical consumers

8 Canadian Ethical Diamonds and Identity Obsession... 165

Very often in movies, like the one with DiCaprio, *Blood Diamond*, they talk about the digging of diamonds in Africa, in the Belgian Congo region, etc. where the working conditions are very bad. From what I have seen from the documentaries, little boys or children worked in conditions that were inhuman, extremely uncomfortable. However, knowing from various readings that ethical diamonds come from Canada, I know… that Canada is an evolved country as regards the care of workers, that it is a western first-world country with laws regarding the protection of labor. So, I imagine Canadian mines as industrial contexts similar to European contexts. (Mattia 43, Entrepreneur, Reggio Emilia)

Extant research suggests that the greater the desire to know about a given topic, the greater the confidence individuals will take in the information they find (Laurin et al., 2010; van der Toorn et al., 2011). Moreover, Shepherd and Kay (2012) argue that in order to resolve an uncomfortable psychological state precipitated by a lack of knowledge on a given topic, people tend to legitimize and increase their trust in information that will make them feel more comfortable. Our own participants appeared to be emotionally ill at ease when their lack of knowledge of CEDs was openly on display. Thus, it is possible to hypothesize that Italian ethical consumers' desire to increase their knowledge about CEDs is rooted in a desire to act in concert with their value system. As a Simone Righi customer reported:

When I went to buy a ring for my wife… the jeweler told me why he started selling ethical diamonds. He explained to me the history of these diamonds and the story about traceability. The reasons why he offers his customers this type of jewelry met my ideals. I felt comfortable with him because we could understand each other. Then, in my case, I wanted to give my wife a gift with a ring on which to mount three diamonds symbolizing our three children. And the idea that these diamonds were also ethical and tracked convinced me 100% to make this purchase. (Luca, 58, Lower Bologna)

As Luca clearly states, the information on traceability in the narrative from the jeweler increased his confidence dramatically, allowing him to immediately reach a purchase decision. The concept of traceability not only allows consumers to differentiate CEDs from other diamonds, it also enables them to connect the various parts of the supply chain in their imaginations by mentally picturing each step of the stone's journey from mine to consumer. In contrast to other stones without certification, CEDs thus easily assume a connotation of identity, as each stage of the supply chain can potentially be made known. CED traceability offers a complete overview of a diamond's journey from mine to customer, which through its transparency connects the end point—the customer purchasing jewelry featuring the diamond—to all those involved along the way, from the workers along the supply chain to cutters, polishers, and setters through to the jewelers. In so doing, CEDs take on unambiguous contours. Schlosser (2013) suggests that the concept of purity in discussing CEDs is related less to the material qualities of the diamond than to a series of characteristics both physical and geographical that evoke an almost pristine Nordic environment (beyond extractive industrialization) in which CEDs are mined (ibid.). Purity is further embodied by the CED itself in accordance with value characteristics describing CEDs as ethically pure, that is, morally correct in terms of worker rights (Remotti, 2010) in the gemstone industry.

8.5 Conclusion and Future Directions

The anthropologist Mary Douglas, in her book *Purity and Danger* (1966), considers the importance of the concepts of purity, pollution, disorder, and danger in constructing social order. Order in society implies certain restrictions; from all materials available, a selection has been made. Disorder, on the other hand, spins patterns into chaos. While that disorder is unlimited, patterns are nonetheless still possible; disorder thus has the potential to be both dangerous and powerful (Douglas, 1966). In this study, we noted that jewelers, particularly in trade fairs and formal contexts, were hesitant to discuss ethical diamonds because doing so would upset the existing order in the jewelry business: to embrace ethics in jewelry production is to accept significant changes in how workers who mine metals and gemstones and fabricate jewelry are treated. Such changes will inevitably increase the costs of production, which may not be offset by higher prices for consumers. The oft-discussed concept, typically in the context of apparel, of purchasing items of higher quality but in lower quantity can result in lower revenue for industry stakeholders.

The potential adjustments in jewelry production and manufacture in response to issues of ethical sustainability, in terms of environmental, economic, and social degradation, is to varying degrees also threatening current views of nature and culture, which are already under siege in a world actively experiencing the long-term impact of the industrial revolution. In response, as issues of sustainability and ethical behavior come ever more sharply into focus, extrapolating ethical questions relating to the production of food, apparel, and home goods, among many others, to include jewelry is a natural extension of such concerns.

In response to the aforementioned issues, consumers we interviewed have a personal stake in sustainability, with their views evolving in tandem with their awareness. While our respondents were aware of issues associated with blood diamonds from various countries in Africa, they had not heard about any unethical conditions for employees in Canada. They are therefore predisposed to accept as truthful statements by the jewelers and to also assume that their own understanding of employee rights in countries such as Canada is correct. To a great extent, they are correct; Canadian labor laws are by any definition far stricter than their counterparts in less developed countries (D'Angelo, 2019) and consumers in our study were well aware that social structures in place in Canada protect workers more than their equivalent in Africa (Le Billon, 2006). We therefore argue that narratives of virtue through consumption can lead consumers to feel they have the ability to buy what they believe in, that opening their wallet is in and of itself an act of power. Questions of authority arise: When consumers rely on experts (e.g., jewelers) for the knowledge to make moral purchase decisions, who will they ultimately consider trustworthy? If only a very few jewelers endorse ethical diamonds, will the value of their customers' ethical purchases fall (a particular sensitive issue, since CEDs are more expensive than non-certified diamonds)? As Douglas (1966) notes, it is not always easy to recognize existing authority. A sense of potential danger and disorder among consumers must be managed before ethical diamonds can become mainstream. Product traceability is one such mechanism that will help in this process.

In agreement with Remotti (2010), we can say that the concept of identity arises in a context of globalization, in which relations between countries, cultures, and people are now complex. Our Italian ethical consumers deployed identity obsession as a psychological tool to hold on to the narrative about traceability, which helped them act in a way that conforms to their ideals. Therefore, ethical participants were able to distinguish CEDs from other diamonds, even though their initial knowledge of the product was insufficient for making such distinctions; they were further limited by their fears of making poor purchase choices.

Remotti argues: "Identity is the affirmation of our essence, or substance (A = A) and the difference is always a relating and comparing ourselves to others (we A are different from B)" (Remotti, 2010, p. 11); we can similarly state, in relation to our investigation, that CEDs are easily identifiable (as they are tracked) and just as easily differentiated from other (untracked) diamonds. In the ethical diamond, therefore, both characteristics of identity and difference coexist, but they do so less from a material point of view (given that a visual comparison of a rough Canadian diamond and an African diamond will likely not yield obvious differences other than to a professional jeweler, as discussed earlier); rather those differences appear in an intangible way. In fact, wearing a piece of jewelry with a CED rather than a non-certified diamond does not allow one to be explicitly recognized as an ethical consumer—any sense of virtue accruing to the consumer is therefore private.

In this chapter, we have addressed two important aspects of traceability narratives that can serve as a starting point for future analysis. On the one hand, we have seen the challenges jewelers face in positioning CEDs in the jewelry sector in Italy, seemingly due to the consequence of a difficult coexistence between jewelry products incorporating different values within the same sector. The concept of CEDs' traceability serves as an identifying trait. On the other hand, we have seen how the relationship between traceability, the knowledge gaps of Italian ethical consumers in relation to MoO, and identity obsession used as a parameter of choice by CEDs consumers reveal an unconditional trust, worthy of further investigation, as Italian ethical consumers make purchase decisions based on information narrated by jewelers.

Acknowledgments The authors gratefully acknowledge receipt of a Social Sciences and Humanities Research council 435-2013–1211. "This project has received funding from the European Union's Horizon 2020 research and innovation program under the Marie Sklodowska-Curie Grant Agreement no. 837190, as well as the European Union emblem".

References

Armano, L., & Joy, A. (2021). Encoding values and practices in ethical jewellery purchasing: A case history of Italian ethical luxury consumption. In M. A. Gardetti & I. Coste-Manière (Eds.), *Sustainable luxury and jewellery*. Springer.
Bailey, M., Bush, S., Miller, R., & Kochen, M. (2016). The role of traceability in transforming seafood governance in the global south. *Environmental Sustainability, 18*, 25–32.
Bauman, Z. (2007). Identity: Conversations with Benedetto Vecchi. *History of Political Thought, 28*(2), 368–371.
Bhattacharya, C. B., Korschun, D., & Sankar, S. (2009). Strengthening stakeholder–company relationships through mutually beneficial corporate social responsibility initiatives. *Journal of Business Ethics, 85*, 257–272.
Bloemer, J., Brijs, K., & Kasper, H. (2009). The CoO-ELM model: A theoretical framework for the cognitive processes underlying country of origin-effects. *European Journal of Marketing, 43*(1/2), 62–89.
Bradu, C., Orquin, J. L., & Thøgersen, J. (2013). The mediated of a traceability label on consumer's willingness to buy the labelled product. *Journal of Business Ethics, 124*, 283–295.
Brucks, M. (1985). The effects of product class knowledge on information search behavior. *Journal of Consumer Research, 12*(1), 1–16.
Brun, A., Caniato, F., Caridi, M., Castelli, C., Miragliotta, G., Ronchi, S., Sianesi, A., & Spina, G. (2008). Logistics and supply chain management in luxury fashion retail: Empirical investigation of Italian firms. *International Journal Production Economy, 114*(2), 554–570.
Chen, M. F., & Huang, C. H. (2013). The impacts of the food traceability system and consumer involvement on consumers' purchase intentions toward fast foods. *Food Control, 33*(2), 313–319.

Coff, C., Barling, D., Korthals, M., & Nielsen, T. (2008). *Ethical traceability and communication food*. Springer Netherlands.

Danner, N., Keller, A., Härtel, S., & Steffan-Dewenter, I. (2017). Honey bee foraging ecology: Season but not landscape diversity shapes the amount and diversity of collected pollen. *PLoS ONE, 12*(8).

D'Angelo, L. (2019). *Lorenzo. Diamanti: Pratiche e stereotipi dell'estrazione mineraria in Sierra Leone*. Meltemi.

Douglas, M. (1966). *Purity and danger: An analysis of the concepts of pollution and taboo*. Routledge.

Du, S., Bhattacharya, C. B., & Sankar, S. (2010). Maximizing business returns to corporate social responsibility (CSR): The role of CSR communication. *International Journal of Management Review, 12*(1), 8–19.

Escalas, J. E. (2004). Narrative processing: Building consumers connections to brands. *Journal of Consumer Psychology, 14*(1–2), 168–180.

Garcia Torres, S., Albareda, L., Rey-Garcia, M., & Seuring, S. (2019). Traceability for sustainability – Literature review and conceptual framework. *Supply Chain Management: An International Journal, 24*(1), 85–106.

Graulau, J. (2008). 'Is mining good for development?': The intellectual history of an unsettled question. *Progress in Development Studies, 8*(2), 129–162.

Gurzawska, A. (2020). Towards responsible and sustainable supply chain – Innovation, multi-stakeholder approach and governance. *Philosophy of Management, 19*, 267–295.

Joy, A., Wang, J., Chan, J., Sherry, T. S., & Cui, J. F. (2014). M(Art)worlds: Consumer perceptions of how luxury Brand Stores become art institutions. *Journal of Retailing, 90*(3), 347–364.

Laurin, K., Shepherd, S., & Kay, A. C. (2010). Restricted emigration, system inescapability, and defense of the status quo: System-justifying consequences of restricted exit opportunities. *Psychological Science, 21*, 1075–1082.

Le Billon, P. (2006). Fatal transactions: Conflict diamonds and the (anti)terrorist consumer. *Antipode, 38*(4), 778–801.

Marin, L., Ruiz, S., & Rubio, A. (2009). The role of identity salience in the effects of corporate social responsibility on consumer behavior. *Journal of Business Ethics, 84*, 65–78.

McManus, C., McMillan, N., Dowe, J., & Bell, J. (2020). Diamonds certify themselves: Multivariate statistical provenance analysis. *Minerals, 10*, 2–12.

Park, C. W., Mothersbaugh, D. L., & Feick, L. (1994). Consumer knowledge assessment. *Journal of Consumer Research, 21*(1), 71–82.

Remotti, F. (2010). *L'ossessione identitaria*. Laterza.

Romani, S., Grappi, S., & Bagozzi, R. P. (2013). Explaining Consumer Reactions to Corporate Social Responsibility: The Role of Gratitude and Altruistic Values. *Journal of Business Ethics, 114*,193–206.

Roth, K. P., & Diamantopoulos, A. (2009). Advancing the country image construct. *Journal of Business Research, 62*(7), 726–740.

Schlosser, K. (2013). Regimes of ethical value? Landscape, race and representation in the Canadian diamond industry. *Antipode, 45*(1), 161–179.

Shepherd, S., & Kay, A. (2012). On the perpetuation of ignorance: System dependence, system justification, and the motivated avoidance of sociopolitical information. *Journal of Personality and Social Psychology, 102*(2), 264–280.

Spinelli, S., Masi, C., Zoboli, G., Prescott, P., & Monteleone, E. (2015). Emotional responses to branded and unbranded foods. *Food Quality and Preference, 42*, 1–11.

Taleb, N. N. (2007). *The black swan: The impact of the highly improbable*. Pinguino.

Tripathi, S. (2010). The influence of the Kimberly process on conflict and natural resource trade in Africa—What can the UN do? In W. Kälin & J. Voyame (Eds.), *International law, conflict and development: The emergence of a holistic approach in international affairs* (pp. 611–626). EBSCO.

van der Toorn, J., Tyler, T. R., & Jost, J. T. (2011). More than fair: Outcome dependence, system justification, and the perceived legitimacy of authority figures. *Journal of Experimental Social Psychology, 47*(1), 127–138.

9

Sustainability Claims in the Luxury Beauty Industry: An Exploratory Study of Consumers' Perceptions and Behaviour

Panayiota J. Alevizou

9.1 Introduction

Luxury has been known to signify status and power and is considered a means for social transformation (Berry, 1994; Vigneron & Johnson, 2004; Seo & Buchanan-Oliver, 2019). The luxury industry experienced a 13% decline in 2020 but a strong rebound is anticipated in post-2021 (Mintel, 2021a). The beauty sector suffered in the first stage of the lockdown and consumers preferred skincare essentials to discretionary beauty, and although in 2021, the *lipstick effect* buoyed spending, prestige sectors may take longer to recover (Mintel, 2021b). During the pandemic, 51% of consumers purchased a beauty product to boost their mood, whereas 42% considered price and value for money as decisive purchase criteria (Mintel, 2021b). The UK has seen a decline in luxury consumption due

P. J. Alevizou (✉)
Management School, The University of Sheffield, Sheffield, UK
e-mail: p.j.alevizou@sheffield.ac.uk

to both the effect of Brexit and the pandemic making it an important market for further research (Mintel, 2021a).

The importance of sustainable development has been well established and companies have been addressing key environmental and social-economic issues and reassessing their values. In addition, consumers are showing growing concerns over social and environmental problems, pollution and their health. This has motivated businesses to communicate their commitment towards sustainability as a means of elevating consumer preference towards their products, which has resulted in overwhelming numbers and types of sustainability claims leading to consumer confusion (Heroux et al., 1988; Alevizou et al., 2015). An increasing number of independent brands, referred to as *industry disrupters*, have come forward with claims such as 'health benefits', 'clean beauty' or 'natural ingredients', causing confusion in the market in terms of the meaning and the 'true cost of clean beauty' (Alevizou, 2021).

However, the luxury beauty sector has been slower to respond compared to the rest of the beauty industry. This may be due to luxury and sustainability having conflicting values, which implies a weak association between the two concepts (Achabou & Dekhili, 2013; Bom et al., 2019; Athwal et al., 2019). In other words, luxury consumption signals, social status, a focus on aesthetics, hedonism and emotional values (see Hirschman & Holbrook, 1982), whereas sustainability encapsulates altruism and a focus on the wider well-being (United Nations, 2021).

In order to address the market confusion, and the wide variety of claims, industry associations, governments and policy makers are working towards establishing clear guidelines for the avoidance of green/clean washing claims. On a regional level, in 2020, Cosmetics Europe revised its guide *Charter and Guiding Principles for Responsible Advertising and Marketing Communications* to better reflect the current challenges and assist consumers in making sustainable choices (Cosmetics Europe, 2020). On a national level, many governments are supporting businesses and consumers. In the UK, for instance, the Competition and Markets Authority (CMA) provides guidance for businesses to understand and comply with the existing obligations under consumer protection law when making environmental claims (CMA, 2021). Accordingly, claims should be truthful and accurate; clear and unambiguous; must not omit important information; comparisons need to be fair, meaningful and consider the full life cycle and finally they should be substantiated. This guidance is in agreement with the ISO 14001 standards when making environmental claims, as well as, other guides dating back to the 1980s and with current international voluntary principles such as the ISEAL (https://www.isealalliance.org/).

Yet, most studies on luxury consumption have focused on apparel and fashion with little attention paid to the beauty industry despite the growing number and variety of sustainability claims (Sharma et al., 2022). This chapter fills this gap by focusing on the luxury beauty sector and its signals of sustainability. To the author's knowledge, there are only a handful of studies in this area from either a cosmetics consumption or a sustainability and luxury perspective. The primary aim and theoretical contribution of this chapter is to join two streams of literature—*sustainability claims* and *luxury beauty consumption*—and improve the understanding of consumer decision making in the beauty luxury industry under the influence of sustainability signals. In conceptualising luxury beauty consumption and sustainability claims engagement, this study undertakes a qualitative approach with two key research questions:

RQ1: *What are consumers' perceptions of luxury beauty brands and their signals of sustainability?*
RQ2: *How are consumers engaging with sustainability claims and how does that process influences their perceptions of 'self'?*

Following the perspective of Ajitha and Sivakumar (2017), in this chapter, luxury beauty brands are defined as cosmetics with limited supply and high price offering the individual the opportunity to personalise their appearance as well as a feeling of self and social desirability in addition to functional values. The term *sustainability claims* signifies environmental, social and economic claims and messages made from companies wishing to signal their approach to sustainable development (Alevizou et al., 2018).

9.2 Conceptualising Sustainability Claims and Luxury Beauty Consumption

Beauty has been the epicentre of both the fashion and the beauty industries for centuries. Historically changes in the external environment seem to affect both industries and as women's fashion choices evolved so did their cosmetics and personal care choices (Matthews, 2018). The beauty industry also referred to as the cosmetics and personal care industry is part of the fast-moving consumer goods (FMCG) industry and is usually divided into five main business segments: toiletries, fragrances, skincare, haircare and make-up (Statista, n.d.). Furthermore, beauty products can be subdivided into premium and mass production segments depending on their marketing mix elements (Statista, n.d.). It must be noted that the structure of the beauty industry is highly regulated and complex (Callaghan, 2019). Yet, existing regulation is inadequate to protect consumers as the enforcement procedures are insufficient (Riccolo, 2021).

The historical origins and elaborations of the idea of luxury can be traced back to the Hellenic, Roman and Christian frameworks (Berry, 1994). The term itself is slippery and presents a number of challenges, as it is frequently used in daily language without a clear understanding or perception of the concept (Wiedmann et al., 2013). Indeed, Kapferer (1997) highlights the complexity behind luxury and luxury brand definitions. The author states that the problem with the word *luxury* is that "it is once a concept (a category), a subjective impression and a polemic term" (Kapferer, 1997, p. 251).

Seo and Buchanan-Oliver (2019) summarise the existing classifications and conceptualisations of brand luxury consumption into two broad perspectives: *Product-centric* (i.e. what is brand luxury) and *consumer-centric* (i.e. how consumers internalise meanings). The authors further discuss *social meanings* and *personalised meanings* under the consumer-centric perspective encapsulating meaning making at social and individualised levels. As previously noted, due to the lack of studies on consumers' perceptions of luxury beauty products and their signals of sustainability, as well as the two streams of literature involved (i.e. sustainability claims and luxury beauty products consumption), the focus of this chapter will remain on both perspectives as outlined by Seo and Buchanan-Oliver (2019).

9.2.1 Luxury Beauty Products and Sustainability Claims

Under the product-centric perspective, a luxury brand becomes a matter of identifying tangible and intangible product attributes and, as such, the assumption that luxury *can be crafted* (Seo & Buchanan-Oliver, 2019).

In the past few years, the beauty sector altogether has been relatively vocal, in terms of sustainability. The term *sustainability* often serves as a sort of *catch-all phrase* to describe an ongoing phenomenon with desirable characteristics that are replicated in the long term, such as a sustainable financial or economic policy or competitive advantage (Borland et al., 2016). Most efforts in the beauty industry fall under the environmental aspect rather than the socioeconomic ones (Bom et al., 2019). For instance, studies investigated the role of sustainability in the cosmetics industry noting that one main challenge is replacing unsustainable ingredients with more sustainable ones (ibid). Indeed, a current debate within the beauty industry is the *natural* versus *synthetic* ingredients with a growing number of companies overemphasising their 'natural ingredients' (Lin et al., 2018), whereas others have gone further with statements of 'free off' ingredients (Grabenhofer, 2020). This approach has created concerns amongst policy makers, and industry stakeholders in terms of product safety (Grabenhofer, 2020). In addition, Secchi et al. (2016) found that an alleged 'natural/eco-friendly' ingredient might result in a less preferable environmental profile from a life cycle assessment perspective.

Sustainability certifications could potentially be seen as a solution to the wide variety of claims (and debates) in the beauty industry as Bom et al. (2019) call for a single certification focused on sustainability, as this is not covered by the existing organic and natural ones. At this point, it should be noted that studies have supported certification standardization since the early 1980s as currently more than 450 labels in 199 countries and across 25 sectors (Ecolabel index, 2021). To make things even more complicated the beauty industry shares a number of certifications with other industries such as the food and fashion industries. For instance, Lin et al. (2018) highlight the effect of the food market and the FMCG sectors on the beauty industry as consumers become more health conscious.

In addition to the conflicting values between the two terms in the luxury sector Achabou and Dekhili (2013) note that despite the increasing sustainability concerns, consumers prioritise product quality and brand reputation over environmental brand commitment and perceive the use of recycled materials in their luxury products negatively. This brings luxury beauty brand communication in a crossroad. In response, De Angelis et al. (2017) call for a reimagining of the two terms in the luxury sector as the concepts share overlapping values and practices. Focusing on the luxury beauty sector and its signals of sustainability is important, as studies have shown that among all luxury items, cosmetics and perfumes are the most widely and frequently consumed products (Dubois & Laurent, 1996).

9.2.2 Consuming Luxury Beauty Products and Sustainability Claims

The evolving portrayals of *beauty* seem to go hand in hand with consumers' quest for the *beautiful* (i.e. object, figure and experience). As such, their consumption is characterised by a certain dynamism and seems to result from them seeking to reinvent themselves and fit in their desired social worlds (Holt, 1995). When consumers do not manage to match their 'actual' self to their 'ideal' or 'social self', they may experience negative emotions (Higgins, 1987). In such cases, consumers may engage in phases of increased fashion consumption (Alevizou et al., 2021). In addition to increased consumption, consumers may turn to luxury brands consumption, as luxury brands are more socially visible and accepted (Vigneron & Johnson, 2004). Bauer et al. (2011) note that consumers value the experiences with luxuries as they are able to generate special private moments. In other words rather than signalling social status, luxury good signal their *private self and experience*. In this aspect, consumers seem to apply an additional emotional element in their decision making (Steinhart et al., 2013).

Studies indicate that consumers are asking for additional information related to inclusion and diversity from their beauty brands (Pounders, 2018). It is not surprising that some studies note that consumers may see a closer fit between luxury and sustainability in the beauty industry (Athwal et al., 2019). A few brands are incorporating social messages, inclusivity, diversity, and promoting self-esteem in their communication platforms. The results in terms of effectiveness of such messages are mixed. For instance, Halliwell and Dittmar (2004) report that it is thinness rather than attractiveness that is an issue for women anxious about their weight, which has been addressed by many fashion and beauty brands in their campaigns. However, some studies report that consumers perceive adverts featuring plus size models as promoting unhealthy behaviours (Pounders, 2018).

In terms of consumers' perceptions of sustainability and luxury product consumption, Beckham and Voyer (2014) highlighted the complexity of consumers' associations between luxury and sustainability consumption. In their study, consumers seem to associate luxury with unsustainability. The authors note that this was not the case when luxury was compared to high street brands. They also note that consumers deemed luxury brands less desirable and luxurious when labelled as sustainable. Overall, they found that consumers experience difficulty associating sustainability with luxury. Similarly, Davies et al. (2012) state that ethical-luxury is unlikely to keep pace with the growth of ethical commodities; however, in their study, consumers stated that they did not think of ethics when shopping for luxury goods. On the contrary, Steinhart et al. (2013) explored the environment claim perceptions for utilitarian and luxury products and found that consumers evaluated more favourably the products with an environmental claim. The authors found that consumers considered the claim as a utilitarian aspect of the product which increased its perceived functionality but also the justification of using the luxury product.

To further explore these inconsistencies a strand of research has classifying consumption/consumers of luxury goods. Dubois and Laurent (1996) explored luxury consumption under the supply–demand factors and identified a third type of luxury consumer which they called 'Excursionists' which in contrast to other types (i.e. no access and permanent access to luxury) their acquisition of luxury items is occasional and under specific circumstances which contrasts to their daily life. From a social practice perspective, Seo and Buchanan-Oliver (2019, p. 418) identified five distinct forms of luxury brand consumption: (1) investing in brand luxury, (2) escaping into/with brands, (3) perpetuating an affluent lifestyle, (4) conveying social status and (5) engaging in self-transformation. The authors note that there is a broad and paradoxical range of personalised meanings consumers construct about luxury brands and consumer engagement with luxury consumption can be situational and contextual. Henninger et al. (2017) focused on luxury consumption amongst Chinese consumers and found four consumption types. In particular, they argue that their category of luxury 'indulgers' ignores the moral compass when they decide to purchase luxury products. Most importantly, the authors note that whilst sustainability is not a key decision making factor, consumers are expecting high corporate social responsibility (CSR) standards from their luxury brands as a minimum requirement but at the same time the authors caution brands when communicating CSR intensively and on all their communication platform due to consumer scepticism. From a consumer segmentation point of view, Makkar and Yap (2018) discussed their typology of inconspicuous consumption and identified four consumer segments (fashion influencers, trendsetters, fashion followers and luxe conservatives). Contrary to previous studies, the authors support the ability to move upwards in social status by both accumulating cultural capital and developing the inner self. However, there are limited/no studies exploring consumers' luxury beauty brand consumption from a sustainability point of view, which is addressed in this chapter.

9.3 Methodology and Analysis

This study adopted a phenomenological approach with the aim of eliciting in-depth information on consumers' perceptions. As the key research questions surrounded two strands of literature and to the best knowledge of the author, there are no previous studies exploring the specific topic a qualitative approach was deemed as the most appropriate (Patton, 2002).

A convenience sampling approach was used where 11 female participants from the ages of 30–50 were recruited who had purchased luxury products (skin care, make-up, fragrance) in the past year. Consumers within these age groups are considered the strongest segments for luxury purchases (Statista, 2021). In addition, participants were asked to take photographs of their purchases as seen in their 'personal spaces'. These visuals were used as an elicitation technique probing interviewees to discuss lived rather than hypothetical experiences. Photo-elicitation has been used as a projective technique and involves showing photographs to participants (either their own or the researcher's) and then asking them to talk about what they see (Barton, 2015; Walker & Widel, 1985). Semi-structured interviews were used as they are appropriate to generate rich data (Cassell, 2015). Data saturation, for the key research themes meaning the point where further data collection does not elicit new information (Patton, 2002), was met after eight interviews. The interviews lasted up to 1 h and were conducted online (Table 9.1).

Due to the amount of visual and textual material, NVivo was used to manage data collection. Ethical approval was granted by the author's institution and participants were provided with an information sheet detailing the research focus as well as the approach. Respondents were recruited via the professional networks of the researcher. The requirement was for participants to have purchased a luxury beauty product within the past year. As such the aim was to capture a wide range of participants (i.e. as classified by Dubois & Laurent, 1996) rather than focus on the most affluent ones. Snowballing recruitment was also adopted and participants recommended their family or friends as potential interviewees.

The interview protocol consisted of three main parts. In the first part of the interview, participants were asked about their perceptions of beauty and luxury beauty and their awareness of sustainability signals. In the main part of the interview, two key themes were explored and were centred on the key research questions. In particular, questions such as the following were asked: "What is your relationship with your luxury beauty brand?" "What do you know about the social and environmental responsibility of your luxury beauty brand?" "What type of sustainability information- communicated by your luxury brand- are you aware of?" "How do you feel about it and what does it means to you?"

Data were analysed following a thematic analysis approach (Braun & Clarke, 2006). This involved: *audio data familiarisation; verbatim transcription; reading the transcripts and keeping initial notes; generating the*

Table 9.1 Participant profiles and purchases

ID	Age group	Profession	Brands	Other luxury categories
Emma	30–40	General practitioner manager	Dior, Estee Lauder, Bobbi Brown	Handbags, shoes, clothes (discount retailers)
Diana	40–50	University professor	No7, Indie luxury cosmetics	Handbags, clothes, accessories, jewellery
Rachel	40–50	Marketing	Indi luxury brands, dermatologist created/approved/ marketed brands	Handbags, shoes, clothes, jewellery
Sophie	40–50	Psychologist	Dior, Estee Lauder, Bobbi Brown	Handbags, shoes, clothes, jewellery, cosmetic procedures, diet/ visits to dietician
Mary	30–40	Teacher	Bobby Brown, Bumble and bumble, Dior	Not mentioned
Elisabeth	40–50	Teacher	Dior, Lancôme	Handbags, shoes, clothes, jewellery
Sherry	40–50	Public sector worker	Dior, Lancôme, dermatologist created/approved/ marketed brands	Cosmetic procedures, diet/ visits dietician/
Donna	40–50	Real estate agent	Indi luxury brands, dermatologist created/approved/ marketed brands	Diet/visits to dietician, Gucci, Prada, Dolce and Gabbana, Hermes
Marta	30–40	Psychologist	Dior, Lancôme, Estee lauder, Bobbi Brown	Handbags, shoes, clothes, jewellery
Anna	30–40	Teacher	Indi luxury brands, dermatologist created/approved/ marketed brands	Handbags
Johnnie	40–50	Marketing	La Mer, Christian Dior, Charlotte Tilbury, Elemis	Handbags, shoes, clothes

initial codes (themes and subthemes); reviewing key emerging themes and creating a thematic representation; following an ongoing analysis and producing a written report.

9.4 Findings and Discussion

When defining luxury beauty products, participants referred to both utilitarian and hedonic brand benefits which agree with previous studies (Apaolaza-Ibanez et al., 2011) as participants seem to voice their preference for functional aspects of the products (i.e. ingredients targeting specific concerns) as well as more aesthetic (packaging and product texture) and sensorial stimuli (scent and feel on the skin). Participants also referred to premium price and quality (Vigneron & Johnson, 2004), the desire for luxury products (Belk et al., 2003), social status signalling (Han et al., 2010) and an escape from their daily routines and *self* (Hemetsberger et al., 2012). As such, consumers' perceptions of their luxury beauty brands and their sustainability signals are discussed from a *product* and a *consumption* perspective which are interwoven and do not seem to exist in *isolation*.

9.4.1 The Luxury Beauty Brand and Its Sustainability Signals: A Product Perspective

The meanings of both luxury and luxury beauty brands found most participants in agreement. For these consumers, luxury brands are connected to higher prices, limited availability, social status signalling and a sense of self-transformation. All participants were quite vocal when discussing their luxury beauty brand purchases and were keen to discuss researching, buying, using, storing, displaying and disposing their preferred luxury beauty products. Most consumers seem to agree that when a luxury beauty product 'works for them' they will keep purchasing it. This was attributed to both the financial and time investment towards these purchases and their *fit* with consumers' personalities and beauty needs/concerns. As Sophie (40–50) stated:

> I'm very loyal [to my luxury brands]...I am satisfied with the brands I am currently using...I mean I tried all...a lot of natural products, as well as, cheaper brands but they did not work for me. So I went back to my regular ones which I will keep buying for life!

In addition, participants referred to the shopping experience itself as a benefit in acquiring luxury beauty brands. As Diana (40–50) mentioned the 'idea' of luxury consumption is connected to her well-being.

> That [spending on luxury beauty] is what makes people feel better about themselves and feel happy and these are the things we do in order to value our self in a way...and be kind to yourself, and if being kind to yourself means buying La Mer or a really expensive face cream go and do it!

Participants also stressed the importance of the retail experience as most participants preferred to visit their favourite retailers rather than purchase luxury brands online. This signals that the luxury beauty experience starts prior to buying/touching the product.

A common response pattern during the interviews was the silence following the question about product/brand sustainability signals. As an initial reaction, most participants were apologetic for not considering sustainability as a factor in their luxury beauty brand purchases. As Rachel (40–50) stated:

> I don't think I have, really...that is something I am conscious about...That's why I go to Luxury Brand X as I know what I am getting, I would like to think that [brand X] products are green and sustainable otherwise they should not be selling them because they are a big company. But I don't know, I haven't really given that an awful lot of thought too.

Most participants stated various reasons for not considering sustainability during their luxury beauty products purchasing. A commonly cited reason was the lack of communication from the part of the brand. As Donna (40–50) mentioned:

9 Sustainability Claims in the Luxury Beauty Industry... 185

No, I am sorry I haven't [considered sustainability]! I follow [brand Y] them on Instagram and the only images I see are those of beautiful people…I do not recall seeing anything about the environment!

Donna continued to discuss her perception for the lack of brand interaction with such messages. In particular, she mentioned that the brand "probably does not prioritise the environment", whereas other participants stressed the focus of their luxury brands on science and being effective rather than *natural* which *is not effective* (Sophie, 40–50). However, most admitted that this was not a primary concern when they purchased these types of products as opposed to food and other fast moving consumer goods. Indeed, as Henninger et al. (2017) pointed out consumers ignore the moral compass when shopping for luxury goods and as McDonald et al. (2012) some consumers seem to be selectors in terms of *their green areas of consumption* as they can be green in one category and not another.

However, after the initial silence—and justification of it—consumers discussed their concerns in terms of the products they purchase (or discontinued purchasing) and their environmental and social considerations.

A few interviewees were concerned about animal welfare, ingredients and testing practices. However, this was not always translated in more research-based and ethical purchase behaviour. Diana (40–50) mentioned that in the past, she used a very effective skincare line which was using animal derived ingredients.

For a while I knew this woman who was doing beauty and she did some beauty for me and she gave me… she got me into trying her cosmetics, and that was in a period of my life where I did the full work, and I did look at the ingredients and they had meres placenta in it …uhm and I was appalled, I had beautiful soft skin and certainly the rituals of cleansing and toning with these products was working but I was appalled and I did not use it again. And of course I did not know at that point that meres placenta is sold on to beauty companies because why would you if you haven't encountered it?

Yet, luxury brands seem to see potential in placenta-derived cosmetics such as equine placenta, which is seen as a superior ingredient for skin care (Lim, 2020). The beauty industry is known to use ingredients (e.g. lactic acid, beeswax, carmine, shark liver oil and vitamin A) derived from animals, gastropods, insects, fish and other species (see PETA, n.d.). This may explain the popularity of "Vegan Approved" certifications. However, none of the interviewees mentioned coming across that certification on their favourite luxury beauty products. This lack of clarity and transparency may induce negative brand perceptions and preferences.

However, this is not always the case as some consumers may decide to consciously lower their ethical standards. An example is Mary (30–40) who kept stressing the importance of animal welfare and testing in her daily consumption but ignored these concerns when purchasing luxury brands "but obviously yes, I try and buy things that are not tested on animals", only to conclude, "I have not checked my Dior lipstick but honestly I think because it is high end …it probably is". It can be seen that for this type of consumers ethical values are lowered, as the assumption is that *luxury and sustainability do not fit within a product* and as such, a choice needs to be made. Mary attributed this behaviour to the lack of trust towards luxury brands and sustainability claims:

> Honestly, the main problem that I have is that I do not trust most of the brands even when they say they are sustainable and ethical. There's… loads of them have the Bunny Mark or saying that they are …but they are not…or they are lying and say 'oh yeah you can recycle all or plastic packaging but actually only three recycling centres in the UK take it and your local council doesn't take it anyway'. So it is kind of less of a concern because it is less achievable.

She also highlighted the effect of global regulatory differences in terms of animal testing as some brands selling in specific parts of the world will need to be tested on animals.

Packaging was mentioned as a key concern as consumers felt more knowledgeable discussing packaging than other aspects of the supply chain and production process. A few consumers purchased refillable skincare products, which seems to be a popular luxury beauty market trend. As Rachel (40–50) stated:

> For me packaging is important as it creates that sense of luxury, so yes packaging is important. On the other hand, I also prefer refillable products as waste is reduced considerably. So yes, I have been using refillable skincare products and I will continue doing so.

This is not surprising as studies seem to stress the priority of packaging and waste reduction for consumers (Mintel, 2020). In addition, reports stress the reduced environmental impact of reusable versus single-use packaging (Zero Waste Europe, 2020). Other consumers did not have similar perceptions, awareness or experience. As Johnnie (40–50) mentioned:

> I have used refills yes, I bought a [brand C] foundation in a reusable tin. I was not happy as the tin broke right after the second refill.

Finally, consumers stated that they had no awareness of any luxury beauty brands certifications or labelling. This is not surprising as luxury beauty is not particularly vocal on product packaging in terms of sustainability nor on its communication. As discussed earlier in the context of other luxury sectors, consumers may perceive the use of recycled materials in their luxury products negatively (Achabou & Dekhili, 2013). In the context of beauty products and especially in luxury beauty a few consumers mentioned that more 'natural-based' products are not as effective as luxury beauty brands due to the investment of luxury brands into science backed-up results. As Jen (30–40) mentioned:

> I am unsure of the effectiveness of natural products, I mean what is in them? When do they expire? Where do you store them? With brands like [luxury Brand F] I can see the ingredients…I know there is an entire scientific team behind the brand and …er…quality controls, and audits and all that… but with natural products …well I doubt they have resources for all that.

This view ties well with the low consumer awareness of what sustainability means for the beauty industry. In other words, some consumers do not perceive a good fit between science and sustainability in the beauty industry. This stereotypical view creates a double challenge for brands in the beauty industry as on the one hand luxury is seen as conflicting with sustainability and on the other hand 'more natural based beauty' as not being backed up by science and effectiveness.

9.4.2 The Luxury Beauty Brand and Its Sustainability Signals: A Consumption Perspective

Consumers discussed usage and disposal of their products. A few consumers pointed out that purchasing luxury cosmetics means purchasing less often and, as such, economising and creating less waste. As Emma (30–40) pointed out

> I think buying luxury beauty products is more sustainable. I noticed that I consume less product, I become more frugal… and I try to squeeze out every last drop out of my products.

Whereas Donna (40–50) points out that when she travels, she never packs her premium skincare but buys cheap skincare at her destination, which she can throw away instead of carrying it back home. Such throwaway consumption behaviour can be encountered in the fashion industry where it has been reported that consumers will purchase "throwaway outfits" (Censuswide, n.d.). The effects of fast fashion are well-documented in the literature and this study highlights a similar consumption behaviour in the beauty industry. This beauty products throwaway trend deserves more research.

Participants also discussed their luxury beauty consumption during lockdown. Most stressed the lack of availability of their brands online and others highlighted some changes in the frequency of consumption during lockdown, as well as, the introduction of more luxury brands in their daily routines. As Johnnie (40–50) mentioned:

I was finding lockdown really hard...and I looked at my face a whole lot more than before lockdown...and I started focusing more on signs of ageing...dark circles...wrinkles and all that. Working from home and being 'connected' did not help either. Zoom was my new mirror... I think it made me more self-conscious and slightly disappointed...and yes...! I ended up buying more luxury skincare products than before...and it was the first time I actually considered more invasive solutions.

Accordingly, consumers seem to connect their well-being with luxury beauty brand consumption. In this study, women seek confidence in their skincare routines and brands in order to improve their perceived desired or ideal self. This is in agreement with previous studies indicating the use of cosmetic products as a relief from feelings of dissatisfaction with the actual self (Apaolaza-Ibanez et al., 2011). In addition, this study highlights that consumers make comparisons not only with attractive role models (see Apaolaza-Ibanez et al., 2011) but also with *past* or *desired* versions of the self.

Finally, participants discussed their spaces of consumption and their skincare routines and referred to them as self-care and well-being *rituals*. When discussing their routines, none of the interviewees mentioned any environmental and social concerns. A few consumers discussed their recycling practice and they seem to be guided by product/package labelling. As previously mentioned, some consumers were sceptical in terms of recycling these products, whereas others admitted not knowing what to do with the empty packages. None of the interviewees was aware of any retailer or brand recycling scheme. The luxury industry advises consumers to (1) research recycling restrictions in their local area, (2) check packaging and labelling guidelines, (3) separate components and (4) clean and condense containers before recycling them (GPA Global, 2021). Yet, none of the participants were aware of these guidelines.

9.5 Conclusion, Contribution, Limitations and Future Research

This study informs current knowledge on luxury beauty consumption and the role of sustainability signals. Even though the sample of this study is its key limitation, the aim of the study was to explore lived experiences and understand the relationship between luxury beauty consumption and sustainability signals. First, the attention is drawn to the *battle of selves'* spectrum of consumer decision making and luxury consumption. In addition to studies emphasising the connection of luxury consumption with, self-concept enhancement (Vigneron & Johnson, 2004), identity and 'extended self' reflections (Belk, 1988), seeking 'uniqueness and self-transformation' (Seo & Buchanan-Oliver, 2019) this study highlights the inner *decision making in terms of values and selves*. As such, the theoretical contribution of this study is the *five emerging behavioural patterns*.

On the one end of the spectrum, the more ***luxury minded self*** is leading beauty consumption but also has never considered any sustainability signals. Sophie (40–50), Johnnie (40–50), Sherry (40–50) and Emma (30–40), for instance, seem aware of sustainability and are consuming more sustainable options in other sectors but have consciously decided to ignore these, when dealing with their beauty routines. For these consumers, the weight of the decision is on the product attributes, effectiveness, experience, skincare needs and the science behind the brands. These consumers are aware of sustainability signals but prefer to ignore them in the name of their beauty rituals. There were no 'apologetic responses nor reactions' but clear statements of priorities in beauty consumption. Very close to this behavioural consumption pattern is the ***luxury minded 'trapped' self*** encapsulates consumers who have strong ethical values in other parts of their lives but do not believe that the industry can be sustainable by default, which makes them sceptical, 'trapped' and distrustful towards beauty brands and sustainability. These consumers seem to be informed and cite greenwashing and other unethical brand practices as a justification for ignoring their *values*.

Next to these consumers are ones that are ***luxury minded but unreached*** by sustainability signals. Elisabeth (40–50) apologised for never giving it a thought or not being aware, but also Donna (40–50) justified her luxury beauty choices as "luxury brands do not want to talk about these things or simply they do not care". In this category, luxury brands are not successful with their sustainability communication (if any) which may present a risk, as these consumers seemed more *apologetic* in their responses and stated that maybe they should do more research in the future. This is not to say that brands should clean/greenwash, but potentially more transparency in terms of their sustainability strategy and practices may be beneficial to their customer base.

Closer to the more **sustainability and luxury minded** consumers are those who have selected more sustainable luxury options for ethical reasons. This pattern of behaviour is more receptive to brand communication and is ready to adopt more sustainable consumption practices.

At the end of the spectrum are consumers who are more **sustainability minded**. These consumers seem to connect their consumption with ethical values and as such, abort luxury brands not fitting their standards. Transparent, clear and substantiated sustainability labelling and brand communication are important elements of decision making as beauty consumption is part of the overall lifestyle. McDonald et al. (2012) report similar patterns in their study with their category of *translators* and this study echoes their findings of this strictly ethical minded category of consumers who will *sacrifice* their beauty rituals in spite of the noted effectiveness.

Finally, it should be noted that these patterns are not rigid expressions of the self(ves) but fluid according to the context or type of beauty product consumed (skincare, makeup, fragrance etc.). This is in agreement with past studies claiming consumer agency and switching behaviour between their luxury consumption practices in order to address contextual and situational concerns (Reckwitz, 2002; Seo & Buchanan-Oliver, 2019; McDonald et al., 2012; Henninger et al., 2017).

9.5.1 Managerial Implications

This study presents some implications for brand managers and marketers. First, the different behavioural patterns signal the complexity of the luxury beauty brand sector. This study echoes previous studies on luxury consumption in fashion (see Henninger et al., 2017; Alevizou et al., 2021) and further highlights particular patterns in the luxury beauty industry. Brand managers, beyond exploring demographics and social status or lifestyle considerations can focus on engaging with consumers' values by addressing them in a sustainable, verifiable, ethical and transparent approach. Most importantly, consumer education in terms of clean beauty claims becomes vital as a wave of *cleanwashing* claims seems to disrupt brands committed to sustainable development. In addition, organisations need to pay more attention to the disposal of their products as consumers seem to be confused as to where to (or whether) recycle their products. Opportunities seem to exist in creating more reusable/refillable products and as such more research needs to be invested in sustainable product design and packaging.

As previously mentioned the sample for this study is a limitation. Future research can explore further these behavioural patterns in terms of consumer perceptions of product efficiency, brand communication, product disposal and sustainability signals.

References

Achabou, M. A., & Dekhili, S. (2013). Luxury and sustainable development: Is there a match? *Journal of Business Research, 66*(10), 1896–1903.

Ajitha, S., & Sivakumar, V. J. (2017). Understanding the effect of personal and social value on attitude and usage behavior of luxury cosmetic brands. *Journal of Retailing and Consumer Services, 39*, 103–113.

Alevizou, P. (2021). Getting creative with sustainability communication in the beauty industry: Exploring on-pack practices and consumers' perceptions. In *Creativity and marketing: The fuel for success*. Emerald Publishing.

Alevizou, P. J., Oates, C. J., & McDonald, S. (2015). The well(s) of knowledge: The decoding of sustainability claims in the UK and in Greece. *Sustainability, 7*(7), 8729–8747.

Alevizou, P. J., Oates, C. J., & McDonald, S. (2018). Signaling sustainability: Approaches to on-pack advertising and consumer responses. In *American Academy of Advertising. Conference* (pp. 74–77). Proceedings (Online).

9 Sustainability Claims in the Luxury Beauty Industry... 193

Alevizou, P., Henninger, C. E., Stokoe, J., & Cheng, R. (2021). The hoarder, the oniomaniac and the fashionista in me: A life histories perspective on self-concept and consumption practices. *Journal of Consumer Behaviour, 20*(4), 913–922. https://doi.org/10.1002/cb.1916

Apaolaza-Ibanez, V., Hartmann, P., Diehl, S., & Terlutter, R. (2011). Women satisfaction with cosmetic brands: The role of dissatisfaction and hedonic brand benefits. *African Journal of Business Management, 5*(3), 792–802.

Athwal, N., Wells, V. K., Carrigan, M., & Henninger, C. E. (2019). Sustainable luxury marketing: A synthesis and research agenda. *International Journal of Management Reviews, 21*(4), 405–426.

Barton, K. C. (2015). Elicitation techniques: Getting people to talk about ideas they don't usually talk about. *Theory & Research in Social Education, 43*(2), 179–205.

Bauer, M., von Wallpach, S., & Hemetsberger, A. (2011). 'My little luxury': A consumer-centred, experiential view. *Marketing Journal of Research and Management, 1*(11), 57–67.

Beckham, D., & Voyer, B. G. (2014). Can sustainability be luxurious? A mixed-method investigation of implicit and explicit attitudes towards sustainable luxury consumption. *Advances in Consumer Research, 42*, 245.

Belk, R. W. (1988). Possessions and the Extended Self. *Journal of Consumer Research, 15*(2), 139–168. http://www.jstor.org/stable/2489522

Belk, R. W., Ger, G., & Askegaard, S. (2003). The fire of desire: A multisited inquiry into consumer passion. *Journal of Consumer Research, 30*(3), 326–351.

Berry, C. J. (1994). *The idea of luxury: A conceptual and historical investigation* (Vol. 30). Cambridge University Press.

Bom, S., Jorge, J., Ribeiro, H. M., & Marto, J. (2019). A step forward on sustainability in the cosmetics industry: A review. *Journal of Cleaner Production, 225*, 270–290.

Borland, H., Ambrosini, V., Lindgreen, A., & Vanhamme, J. (2016). Building theory at the intersection of ecological sustainability and strategic management. *Journal of Business Ethics, 135*(2), 293–307.

Braun, V., & Clarke, V. (2006). Using thematic analysis in psychology. *Qualitative Research in Psychology, 3*(2), 77–101.

Callaghan, T. (2019, September). *A closer look at cosmetics claims* (online). Cosmetics & Toiletries. Retrieved from https://www.cosmeticsandtoiletries.com/regulations/claims-labeling/news/21842939/callaghan-consulting-international-a-closer-look-at-cosmetic-claims

Cassell, C. (2015). *Conducting research interviews for business and management students*. Sage.

Censuswide. (n.d.). *Barley and Barnardo's: The Fast Fashion Crisis*. Available online https://censuswide.com/censuswideprojects/barley-and-barnardos-the-fast-fashion-crisis-research/

CMA. (2021). *Making environmental claims on goods and services*. Retrieved November 2021, from https://www.gov.uk/government/publications/green-claims-code-making-environmental-claims/environmental-claims-on-goods-and-services

Cosmetics Europe. (2020). *Annual report 2020*. Retrieved November 2021, from https://www.cosmeticseurope.eu/

Davies, I. A., Lee, Z., & Ahonkhai, I. (2012). Do consumers care about ethical-luxury? *Journal of Business Ethics, 106*(1), 37–51.

De Angelis, M., Adıgüzel, F., & Amatulli, C. (2017). The role of design similarity in consumers' evaluation of new green products: An investigation of luxury fashion brands. *Journal of Cleaner Production, 141*, 1515–1527.

Dubois, B., & Laurent, G. (1996). *The functions of luxury: A situational approach to excursionism*. ACR North American Advances.

Ecolabel index. (2021). Retrieved from https://www.ecolabelindex.com/

GPA Global. (2021). *How to recycle cosmetics packaging*. Available online https://www.luxurypackaging.co.uk/en/news/2021/apr/22/how-recycle-cosmeticspackaging/

Grabenhofer, R. (2020, January). *The reality of free-from beauty*. (online) Cosmetics & Toiletries. Retrieved from https://www.gcimagazine.com/ingredients/launches-claims/article/21848957/the-reality-of-free-from-beauty

Halliwell, E., & Dittmar, H. (2004). Does size matter? The impact of model's body size on women's body-focused anxiety and advertising effectiveness. *Journal of Social and Clinical Psychology, 23*(1), 104–122.

Hemetsberger, A., von Wallpach, S., & Bauer, M. (2012). 'Because I'm Worth It': Luxury and the Construction of Consumers' Selves. *Advances in Consumer Research, 40*, 483–489.

Han, Y. J., Nunes, J. C., & Drèze, X. (2010). Signaling status with luxury goods: The role of brand prominence. *Journal of Marketing, 74*(4), 15–30.

Henninger, C. E., Alevizou, P. J., Tan, J., Huang, Q., & Ryding, D. (2017). Consumption strategies and motivations of Chinese consumers: The case of UK sustainable luxury fashion. *Journal of Fashion Marketing and Management: An International Journal, 21*(3), 419–434.

Heroux, L., Laroch, M., & McGown, K. L. (1988). Consumer product label information processing: An experiment involving time pressure and distraction. *Journal of Economic Psychology, 9*(2), 195–214.
Higgins, E. T. (1987). Self-discrepancy: A theory relating self and affect. *Psychological Review, 94*(3), 319.
Hirschman, E. C., & Holbrook, M. B. (1982). Hedonic consumption: Emerging concepts, methods and propositions. *Journal of Marketing, 46*(3), 92–101.
Holt, D. B. (1995). How consumers consume: A typology of consumption practices. *Journal of Consumer Research, 22*(1), 1–16.
Kapferer, J. N. (1997). Managing luxury brands. *Journal of Brand Management, 4*(4), 251–259.
Lin, Y., Yang, S., Hanifah, H., & Iqbal, Q. (2018). An exploratory study of consumer attitudes toward green cosmetics in the UK market. *Administrative Sciences, 8*(4), 71.
Lim, A. (2020). Regenerative beauty: Aussie skin care brand sees lucury potential in horse placenta. *CosmeticsDesign-Asia*. Available from https://www.cosmeticsdesign-asia.com/Article/2020/01/02/Aussie-skin-care-brand-sees-luxurypotential-in-horse-placenta
Makkar, M., & Yap, S.-F. (2018). Emotional experiences behind the pursuit of inconspicuous luxury. *Journal of Retailing and Consumer Services, 44*, 222–234. https://doi.org/10.1016/j.jretconser.2018.07.001
Matthews, M. (2018). *A Victorian Lady's guide to fashion and beauty*. Pen and Sword.
McDonald, S., Oates, C. J., Alevizou, P. J., Young, C. W., & Hwang, K. (2012). Individual strategies for sustainable consumption. *Journal of Marketing Management, 28*(3–4), 445–468. https://doi.org/10.1080/0267257X.2012.658839
Mintel. (2020). *UK household care packaging trends: Inc impact of COVID-19 market report 2020*. Mintel.
Mintel. (2021a). *Luxury goods retailing – International*. Mintel.
Mintel. (2021b). *UK COVID-19 and BPC: A year on market report*. Mintel.
Patton, M. Q. (2002). *Qualitative research & evaluation methods* (3rd ed.). Sage.
PETA. (n.d.). *Animal-Derived Ingredients List*. Avialable online https://www.peta.org/living/food/animal-ingredients-list/
Pounders, K. (2018). Are portrayals of female beauty in advertising finally changing? *Journal of Advertising Research, 58*(2), 133–137.
Reckwitz, A. (2002). Toward a theory of social practices: A development in culturalist theorizing. *European Journal of Social Theory, 5*(2), 243–263.

Riccolo, A. (2021). The lack of regulation in preventing greenwashing of cosmetics in the US. *Journal of Legislation, 47*(1), 133.

Secchi, M., Castellani, V., Collina, E., Mirabella, N., & Sala, S. (2016). Assessing eco-innovations in green chemistry: Life cycle assessment (LCA) of a cosmetic product with a bio-based ingredient. *Journal of Cleaner Production, 129*, 269–281.

Seo, Y., & Buchanan-Oliver, M. (2019). Constructing a typology of luxury brand consumption practices. *Journal of Business Research, 99*, 414–421.

Sharma, A., Soni, M., Borah, S. B., & Haque, T. (2022). From silos to synergies: A systematic review of luxury in marketing research. *Journal of Business Research, 139*, 893–907.

Statista. (n.d.). *Cosmetics and personal care*. Retrieved November 2021, from https://www.statista.com/markets/415/topic/467/cosmetics-personal-care/#overview

Statista. (2021). *Luxury shopping in the UK 2021*. Available from https://www.statista.com/study/92163/luxury-shopping-in-the-uk-2021

Steinhart, Y., Ayalon, O., & Puterman, H. (2013). The effect of an environmental claim on consumers' perceptions about luxury and utilitarian products. *Journal of Cleaner Production, 53*, 277–286.

United Nations (UN). (2021). *Sustainable development goals*. UN. Retrieved December 01, 2021, from https://sustainabledevelopment.un.org/?menu=1300

Vigneron, F., & Johnson, L. W. (2004). Measuring perceptions of brand luxury. *Journal of Brand Management, 11*(6), 484–506.

Walker, R., & Widel, J. (1985). Using photographs in a discipline of words. In R. G. Burgess (Ed.), *Field methods in the study of education* (pp. 191–216). Philadelphia, PA: Falmer.

Wiedmann, K. P., Hennigs, N., Klarmann, C., & Behrens, S. (2013). Creating multi-sensory experiences in luxury marketing. *Marketing Review St. Gallen, 30*(6), 60–69.

Zero Waste Europe. (2020). Reusable VS single-use packaging: A review of environmental impact. Available online at https://zerowasteeurope.eu/library/reusable-vssingle-use-packaging-a-review-of-environmental-impact/

10

What Do You Think? Investigating How Consumers Perceive Luxury Fashion Brand's Eco-labelling Strategy

Shuchan Luo, Aurelie Le Normand, Marta Blazquez, and Claudia E. Henninger

10.1 Introduction

In the twenty-first century, consumers are increasingly willing to buy more eco-friendly products (Chen & Chang, 2013). Especially after COVID-19, luxury consumers are increasingly expecting luxury products to be more environmentally responsible (D'Arpizio et al., 2020). This is further enhanced through Gen Y. They not only are anticipated to make up two-thirds of global luxury purchase by 2025 (D'Arpizio et al., 2021) but also have an appetite for luxury products that are classified as sustainable (Granskog et al., 2020). The growing consumer demand from Gen Y has led luxury companies to increasingly integrate a solid environmental focus within their sustainability initiatives. Another reason for luxury companies to strengthen their sustainable practices is that they

---

S. Luo (✉) • A. Le Normand • M. Blazquez • C. E. Henninger
Department of Materials, The University of Manchester, Manchester, UK
e-mail: Shuchan.Luo@manchester.ac.uk; aurelie.lenormand@manchester.ac.uk; marta.blazquezcano@manchester.ac.uk; claudia.henninger@manchester.ac.uk

© The Author(s), under exclusive license to Springer Nature Switzerland AG 2022
C. E. Henninger, N. K. Athwal (eds.), *Sustainable Luxury*, Palgrave Advances in Luxury, https://doi.org/10.1007/978-3-031-06928-4_10

face criticism from the public and (non-)governmental organisations for their unsustainable behaviour (e.g. burning unsold stock) (Cheah et al., 2016). This situation was further reinforced during the pandemic (Amed et al., 2020).

While luxury companies seek to improve their sustainability performance, most consumers see sustainable innovations as essentially commercial and remain sceptic about the authenticity of these corporate statements (Achabou & Dekhili, 2013; Wells et al., 2021). Ivan et al. (2016) suggest different means of communicating sustainability, such as an eco-labelling strategy. Eco-labelling plays a crucial role in conveying corporate sustainability performance to consumers (Osburg et al., 2017). Research by Selfridges shows that almost three quarters of consumers are willing to see relevant certified labels on products (Arnett, 2019). Hence, eco-labelling provides a quality assurance in conveying information about products with environmentally friendly criteria (Bratt et al., 2011; Taufique et al., 2019). Yet, eco-labels have been criticised, as it is difficult for consumers to know what each label entails and/or communicates (Turunen & Halme, 2021). It is, therefore, required to obtain consumer insights on how they feel about eco-labelling and what they mean to them (Taufique et al., 2019).

Past research has generated interest in studying eco-labelling in the luxury industry, yet research from a consumer's perspective is still lacking (Kunz et al., 2020; Osburg et al., 2021). This study addresses this research gap to explore Gen Y consumers' insights into eco-labelling as a sustainability communication strategy in the luxury industry. There are two research questions which are leading the study:

* How do Gen Y consumers perceive eco-labelling strategies in the luxury industry?
* What is Gen Y consumers' understanding about eco-labels used by luxury companies?

10.2 Literature Review

10.2.1 Eco-labels and Eco-labelling Strategies

Consumers increasingly pay attention to products' sustainability attributes when searching for product information (Sharma & Kushwaha, 2019). Rahbar and Wahid (2011) found that consumers can easily switch between brands, especially, when one has a stronger environmental attribute, which may influence a consumer's actual purchase behaviour. Effective communication of sustainability messages can affect consumers' attitudes to actual purchasing behaviour (Turunen & Halme, 2021). However, for consumers, Bratt et al. (2011) highlight, it is difficult and/or confusing to determine the extended quality assessment of products (here—eco-friendly attribution). To reduce consumers' confusion and address credibility issues, existing research suggests that companies could implement an eco-labelling strategy to signal to consumers that what they are doing is more sustainable (Atkinson & Rosenthal, 2014; Sharma & Kushwaha, 2019). Providing information through, for example, labelling is key to reassuring consumers who are increasingly sensitive to brands with environmental messages (Bartiaux, 2008). Clemenz (2010) asserts that claiming to produce eco-friendly products can provide a profitable business strategy for companies, yet needs to be carefully executed in order not to be accused of greenwashing. Greenwashing implies making wrongful claims about environmental credentials. Eco-labelling has become a valuable strategy for communicating environmental information to consumers, as they are verified third-party accreditations (Delmas et al., 2013).

As a business strategy, eco-labelling is designed to fulfil a company's environmental practices, maintain positive public perceptions and strengthen a business' credibility (Delmas et al., 2013; Kearney, 2014). The Global Ecolabelling Network (2021) highlights that eco-labels determine the environmental and/or social characteristics of products within a given product category by considering the life cycle of production. Eco-labels come in many forms, including government-mandated labels (e.g. European Energy label) and self-declarations by retailers or brands (e.g.

stating recycled materials are used, or certain chemicals are not used in production) (D'Souza, 2004). Bougherara and Combris (2009) state that the aim of eco-labelling is 'to internalise the external effects on the environment of the production, consumption and disposal of products' (p. 321). Thus, an eco-labelling strategy is defined as overcoming uncertainty about products' environmental attributes and enhancing the credibility of companies' environmental claims (Bucklow et al., 2017; Taufique et al., 2019). While assuring consumers of the authenticity of the statements, an eco-labelling strategy is to present the certification mark or seal of approval to consumers to indicate the environmental quality of their products or services (Atkinson & Rosenthal, 2014). Turunen and Halme (2021) stress that as a strategy in consumer-oriented sustainability communication, eco-labelling is primarily about bringing sustainability information to the consumer market to support the consumer's decision-making process.

A key challenge however is that there are over 400 eco-labels currently implemented across different industries (Ecolabel Index, 2021). This implies that research on eco-labelling covers different industries, including the oil, fashion and hospitality industry. However, the fashion industry, and the luxury sector in particular, remains an under-researched area to be explored in this regard (Kunz et al., 2020).

10.2.2 Eco-labelling in the Luxury Industry

The International Organisation of Standardisation (ISO) 14020 family provides a set of 'internationally recognized and agreed benchmarks against which they can prepare their own environmental labels, claims and declarations' (ISO, 2019, p. 5), including general principles (ISO 14020), eco-labelling schemes (ISO 14024), self-declared environmental claims (ISO 14021), life cycle data declarations (ISO 14025), footprint Communication (ISO 14026) and product category rules (ISO 14027). Within the fashion context the ISO Type I, II and III categories have been used (e.g. Henninger, 2015; Osburg et al., 2017; Turunen & Halme, 2021), all of which outline eco-friendly properties of a product/service. Examples of labels falling into these categories are:

- Type I: Nordic Swan, GOTS, EU's Eco-label
- Type II: Recyclable content (Plastic Recycle Triangle), Biodegradable
- Type III: Eco-leaf

In the luxury fashion industry, several accreditations are reinforced by luxury brands, such as Positive Luxury (Positive Butterfly Mark, as PBM) and GCC (Green Carpet Challenge) Brandmark (Ivan et al., 2016). The former represents the eco-friendly character of the brand in terms of raw material sourcing, product manufacturing and service marketing (Ivan et al., 2016), which has been implemented by several well-known luxury brands (e.g. Louis Vuitton, Christian Dior, and Givenchy) (Positive Luxury, 2021). The later label (GCC) is in partnership with Eco-age, meaning that rewarded products are made from zero-deforestation and triggered a big awareness and responsibility for a brand's long-term sustainable future goal (Ivan et al., 2016). Gucci and Stella McCartney have been awarded the GCC Brandmark (Eco-Age, 2021).

These labels provide verified third-party information to consumers who may be concerned about various issues at different stages of the life cycle of production, as they focus on different areas of sustainability (Bratt et al., 2011; Henninger, 2015). Yet, the varying aspects of information from eco-labelling may foster more consumer confusion and mistrust (Brécard, 2014). It is because eco-labels in the fashion industry cover a wide range of areas, including agriculture, fibre content and processing (Bucklow et al., 2017). There is currently no uniform eco-label to address consumer concerns about the authenticity of information about products' environmental footprint or whether it is greenwashing by the company (Henninger, 2015; Turunen & Halme, 2021), especially in the luxury sector (Athwal et al., 2019; Davies et al., 2020). The complexity of eco-labelling hinders consumers' ability to understand and trust the information conveyed through eco-labels and to consider more environmentally friendly purchase decisions (Taufique et al., 2019).

Although eco-labelling strategies have various challenges, eco-labels play a critical for quality assurance role in communicating sustainability (Bratt et al., 2011; Atkinson & Rosenthal, 2014). To explain, eco-labelling reduces the information asymmetry between companies and consumers regarding sustainability and maintains further intangible information at the point-of-sale (Atkinson & Rosenthal, 2014; Davies et al., 2020). Eco-labels are an important tool for identifying the sustainable practices and performance of fashion companies (Henninger, 2015). In the luxury industry, Ivan et al. (2016) also stress that eco-labelling is a communication strategy that sensitises consumers to brands that promise positive sustainability performance, thereby increasing the value of consumer consumption of sustainable luxury products. Therefore, this study follows the above-mentioned definition of eco-labelling (Bucklow et al., 2017; Taufique et al., 2019) and the identification from Ivan et al. (2016). The definition of eco-labelling thus on the one hand refers to standards of providing verified sustainability information to consumers, and on the other hand it is a communication strategy for luxury companies.

Despite the literature highlighting the opportunities and challenges of eco-labelling and its associated strategies, it remains under-researched in the luxury fashion sector (Kunz et al., 2020). Currently there is a lack of research to provide a clear understanding of the impact of eco-labelling strategies on consumers' understanding and feelings towards them (Davies et al., 2020; Turunen & Halme, 2021). As a result, this study, thus, chooses consumer perceptions as the research object to look at eco-labelling strategy within the luxury fashion industry.

10.3 Methodology

This study utilises a qualitative approach to explore consumer perceptions of the eco-labelling strategies undertaken by luxury fashion brands. For this study, 10 semi-structured interviews were conducted to gain insights into how eco-labelling strategies are perceived and understood by Gen Y consumers. Conducting semi-structured interviews offers the researcher with a flexible way to structure the interview process, allowing the research to not only follow a pre-determined structure of interview questions but also to adopt questions based on the interviewees' responses (Bryman, 2016). In screening the criteria and characteristics of the participants, purposive sampling was implemented to facilitate the selection of target participants to obtain answers to the research questions (Malhotra et al., 2012).

Gen Y is the primary luxury market force in the coming years and is highly encouraged to develop sustainability in the luxury industry. By considering the participants' actual ages, lifestyles and educational background, this study narrows the general Gen Y (1980–2000) consumers into late Gen Y (1990–2000). Consumers falling into the late Gen Y category are those that are seen to care about the environment and are easily inspired by digital communication (Göbbel et al., 2019; Kuleto et al., 2021; D'Arpizio et al., 2021). Moreover, late Gen Y consumers have a higher level of understanding of sustainability than early Gen Y, as they have been educated in relevant courses during undergraduate/postgraduate (e.g. Sidiropoulos, 2014; Kuleto et al., 2021). This may have an impact on how they receive the relevant information on sustainability (Kuleto et al., 2021). This research thus focuses on late Gen Y consumers.

In this research, participants have been anonymised (Table 10.1). Prior studies have highlighted that previous consumption experience, of here eco-labelled luxury products, could affect their perception of environmental-related information (Rahbar & Wahid, 2011). Based on the frequency of visits to luxury companies' websites/physical stores, participants are classified by level of their experience in browsing companies' websites (High/Low). Participants who visit luxury brands' websites or physical stores on a weekly/monthly basis are identified as high experience consumers. Low experience consumers represent those participants who visit the shop only two to three times per year.

Interview questions are focused on understanding how participants perceive those eco-labels employed explicitly in the luxury industry and what are their understanding and expectation of seeing those eco-labels. The average interview duration is 70 min.

This study employed a grounded analysis technique to gain insights into how consumers perceive and understand eco-labels. Under the grounded analysis, researchers can develop new insights from the analysis by employing a systematic seven-step procedure (*familiarisation, reflection, open coding, conceptualisation, focused re-coding, linking and re-revaluation*) (Easterby-Smith et al., 2015). This method offers a systematic analysis procedure for qualitative research where researchers can efficiently open the way to process the data analysis and keep the consistency of analysis until they obtain a relevant insight (Easterby-Smith et al.,

Table 10.1 Overview of participants

Initials	P1	P2	P3	P4	P5	P6	P7	P8	P9	P10
Gender	F	M	F	M	F	F	F	F	M	F
Experience	HE	LE	LE	HE	LE	HE	LE	HE	LE	HE
Duration	1 h	1 h	1 h	1 h	1 h	1 h	1 h	1 h	1 h	2 h
	24 min	6 min	16 min	28 min	23 min	4 min	15 min	15 min	20 min	10 min

F = Female, M = Male
HE high experience; *LE* low experience

2015). More importantly, this method has also been adopted in previous studies about eco-labels in the fashion industry (Henninger, 2015). This study thus implements grounded analysis for analysing the data.

10.4 Findings

The findings are divided into three categories: *Awareness of eco-labels*, *Diversity of perceptions* and *Expectation of label design*.

10.4.1 Awareness of Eco-labels

In line with previous studies, the results highlight that participants were aware of several eco-labels implemented in the fashion industry, yet, awareness was generally low (Henninger, 2015; Osburg et al., 2017). In this study, data show that most participants were aware of eco-labels adopted by other fashion segments (e.g. fast fashion, sports brands). But no one noticed any eco-labels within the luxury fashion segment. A participant states: '*I didn't see any eco-labels conducted in luxury fashion. But I merely acknowledge those eco-labels implied in fast fashion like Uniqlo, who has utilised Organic Cotton Labels on their garments*' (P2). Here, the quote illustrates participants' engagement with eco-labels in the fashion industry, which is consistent with a prior study saying young consumers proactively address their awareness of fashion sustainability issues (Han et al., 2017). Nevertheless, it also can be seen that the awareness of luxury companies' eco-labels may be lacking.

Additionally, findings highlight that participants were more familiar with eco-labels they saw in the food or tourism industry. As exampled in Sect. 11.2, participants were asked if they were aware of specific labels in the luxury fashion industry, such as the Butterfly Mark. Interestingly, none of the participants knew and/or recognised those labels. Participants have shown a higher consciousness of eco-labels in the food industry than in luxury fashion, such as 'Organic', 'Eco-', or the Recycle Triangle label. A note from participant shows *'I've seen labels like "Organic Food" when I'm grocery shopping […] I think I Am very familiar with this label because*

I see it in my daily grocery shopping or anywhere from the street or at home. For luxury brands, I think they can use this as a reference to effectively communicate their eco-labels to consumers' (P1).

From this participant's statement, one of the reasons for the lack of awareness of eco-labels in the luxury fashion industry is their low presence in this specific sector. Echoing previous studies, eco-labels lack awareness from consumers in the fashion industry (Henninger, 2015; Osburg et al., 2017). One of the novelties from here is that, while participants are aware of eco-labels in the fashion industry, *none of* them were aware of any eco-labels[1] in the luxury fashion industry. Hence, for fashion consumers, the luxury sector's eco-labels may not be as compelling as fast fashion.

10.4.2 Diversity of Perceptions

The suspicion of authenticity about sustainability information hinders communication in the luxury fashion industry (Achabou & Dekhili, 2013; Atkinson & Rosenthal, 2014). Turunen and Halme (2021) stress that transparent and understandable information about sustainability, in which eco-labels act as promoters, may help support consumer decision-making. Here, this section demonstrates diverse consumers' perceptions of eco-labels.

Due to no participants having prior knowledge about eco-labels in the luxury industry, participants are first introduced to some relevant eco-labels (e.g. Butterfly Mark, GOTS). By asking their initial perceptions, they perceive eco-labels diversely, including *Acceptance, Rejection and Unconscious.*

10.4.2.1 Acceptance vs. Rejection

One of the distinct findings here is the initial brand trust, which may affect consumers' perceptions of eco-labels. Consumers with a closer

---

[1] Please note that eco-labels as a term is used loosely in this chapter to include certifications, such as the Butterfly mark.

bond with luxury brands easily accept eco-labels compared to consumers with limited experience with luxury brands. This research affirms the assumption from a previous study (Janssen et al., 2014) that there would be a different perception of sustainability-related information due to consumers' prior knowledge about the brand. Here, results expose a different perception of eco-labels that could be opposite perceptions (*acceptance vs. rejection*). Meanwhile, consumers' prior knowledge about eco-labels might come from their shopping experience.

To explore how consumers perceive eco-labels, interviewees showed that consumers with higher level of shopping experience can positively embrace eco-labels and unconditionally trust the information from luxury brands (P1, P4 and P8). Participants express the reasons for their acceptance from an objective and a subjective perspective, as summarised in Table 10.2. The table has presented those different contents cultivated from objective perspective to subjective perspective. Two categories have been conducted based on how consumers perceived the influence of eco-labels for the luxury brands. It represents that consumers perceive eco-labelling strategies on whether they could affect brands' objective elements (e.g. Brand Reputation, History) or subjective elements (e.g. Social pressure, Brand Trust). The following section will demonstrate how both objective and subjective expressions are accomplished from consumers' perspective.

For instance, '*I believe luxury brands, as a big company with a high social **reputation** and long brand **history**, would not lie on this. Also, they have been watched by the **public** consistently. If they lie on it would be a big damage for the company*' (P8, HE). '*I indeed trust the luxury brands which I like, this brand loves to reinforce me having faith unconditionally to **trust** them and consistently purchase products from them*' (P4, HE).

The findings concur with prior studies (Han et al., 2017; Anido Freire & Loussaïef, 2018), in which a brand's reputation influences consumers'

Table 10.2 Consumers' perspective about the influence of eco-labels (objective vs. subjective)

	Objective	Subjective
Contents	*Brand reputation; brand history*	*Social pressure; Brand Trust*

attitudes about fashion brands' characteristics with sustainability. Here, it confirms that brand reputation can affect consumers' acceptance of eco-labels as brands' characteristics of sustainability in the luxury fashion industry. Results also expose that brand history, social pressure and brand trust may influence consumer acceptance of eco-labels in the luxury fashion industry. Meanwhile, the study can be referred to prior research (Platania et al., 2019); the antecedents of brand love can positively influence the brand's sustainability commitment towards the brand, which manifests as high shopping experience with luxury brands.

However, there were also other opinions, for example P5 (LE) indicates that

> I don't believe any eco-labels that luxury brands in general actively convey. It is a marketing strategy that luxury brands try to convince or even force their consumers to believe them. And I saw some scandal about luxury brands are doing bad for our environment. I think if the brand is good as they are announcing, they don't need any extra tools to present their characteristics.

In line with Anido Freire and Loussaïef (2018), eco-labels may hinder trustworthiness about sustainability attributes from brands in the luxury fashion industry. The reported scandal about sustainability might affect consumers' perception of the luxury brand's fit and sustainability (Janssen et al., 2014). In the luxury fashion industry, the existing brand reputation of sustainability may negatively convey eco-labels to consumers in the luxury fashion industry.

10.4.2.2 Unconscious

There is another voice about employing eco-labels in the luxury fashion industry. Participant P9 stresses,

> I didn't notice and don't think it is necessary for luxury companies to have eco-labels. I don't think eco-labels can identify the quality. I would research the product before purchasing. So I think eco-label will not trigger anything consequences for me.

Having eco-labels might not have any impact on this consumer's perception. Although prior studies (Osburg et al., 2017; Anido Freire & Loussaïef, 2018) stressed eco-labels could affect consumers' attitudes (positively/negatively), here, to the best of the researchers' knowledge, consumers' unconscious perception of eco-labels is new to the literature. Specifying the eco-labels unconsciousness means that consumers do not care whether luxury brands have eco-labels or not. This consumers' perceptions of the luxury brands are unaffiliated, which will not be affected by eco-labels.

In summary, the findings here are consistent with prior literature (Gözükara & Çolakoğlu, 2016), in which Gen Y consists of highly diversified members, who have an interest in brand innovations. This study looked at eco-labels as one of the innovations within sustainability communication in the luxury fashion industry. This generation has not only either positive or negative attitudes of eco-labels but also unconscious perception. It is novel in the literature that consumers may have acceptance, rejection and unconscious perception of eco-labels. It brings a suggestion for further investigations which could try to generalise the findings.

10.4.3 Expectation of Label Design

Results show that consumers actively reflect on eco-labels in terms of how eco-labels are *designed and transmitted*, referring to the assumption from Taufique et al. (2014), who note the design and display of the eco-labels have the potential to influence consumers' attention about the eco-label. This study looks at how eco-labels can be more noticeable from consumers' expectations in response to this assumption.

10.4.3.1 Design and Visibility

Following the statement from prior study (Taufique et al., 2014), 'design and visibility' refers to the look and appearance of eco-labelling, enabling eco-labels to attract consumers' attention.

Consistent with prior studies (Pedersen & Neergaard, 2006; Beretti et al., 2009), results show that the design and visibility of the logo/pictorial of the labels are more attractive than contextual/alphabeted labels for Gen Y consumers. They also expect that the design of eco-labels can incorporate the characteristics of the luxury brands. For instance, one participant suggests that luxury brands can input their iconic design of the products/company combining with sustainable attributes.

> I think the eco-label needs to be designed to integrate both environmental friendly criteria and the brand image. Like Hermes, if they will design eco-labels, I think they need to think about how to combine their signature orange colour of the brand with eco-feature. (P5)

Participants noted that they are more engaged with the logo/pictorial labels than contextual/informed labels, expecting the labels to be designed within eco-friendly features and the brand's initials. Here, a brand's initials refer to the most iconic characteristics of the brand, including the brand's style and product/packaging colour. The study verifies that having the environmentally friendly feature is the base of eco-label design (Turunen & Halme, 2021) and exposes that integrating with a brand's iconic attributes is expected from luxury fashion consumers.

This study found that the design and visibility of eco-labels need to be straightforward rather than complex. Participants point out that they are more willing to see simple, detailed eco-labels with less context, and they easily accept these eco-labels. Referring to a study in the hotel industry, Font et al. (2012) illustrate the intricate design of the eco-labels that facilitates the ignorance from consumers. Consequently, the design of eco-labels could reflect the eco-friendly characteristics of the items and combine the brand's iconic attributes and concise details.

10.4.3.2 Transmissive Methods

The research found that consumers also have different expectations of how eco-labels can be conveyed to them. To the best of the researchers' knowledge, it is still an under-researched area in sustainability communication in the luxury fashion industry regarding the methods of conveying eco-label messages. The results show that consumers expect eco-labels to either be attached to sustainable products or packaging. Like one participant expresses, '*I think luxury brands can try to attach the eco-labels directly on their product. Like Nike has attached specific labels on their special sustainable collection products' surface. And they also have utilised specific packaging for those collections. I think those ways is effectively noticeable*' (P1).

Here, the finding demonstrates that eco-labels are suggested to be communicated visually on products' surfaces or packaging from consumers. The finding approves Testa et al.'s (2015) study that attached eco-labels on the packaging effectively gather consumers' attention about corporate sustainability performance. Furthermore, the study appends that the transmissive methods of conveying eco-labels can be labelled on sustainable products themselves.

In summary, the results of this study can be illustrated within several sections. Firstly, a key contribution of this chapter is to clarify that there is a misinterpretation of communicating eco-labels with luxury consumers. Due to the misunderstanding, consumers could either positively or negatively perceive eco-labels conducted by luxury brands. The research found that eco-labels still lack awareness in the fashion industry referred to in prior studies. Especially in the luxury sector, the investigation initially found that no participants have noticed eco-labels, and participants are more familiar with eco-labels appearing in other fashion sectors (e.g. fast fashion, sporty) than in the luxury fashion sector. Nevertheless, this study is qualitative in nature; the size of the sample is the limitation. There is a suggestion for future research to investigate a bigger simple size or employ an alternative research method to testify further.

Moreover, this study found diverse perceptions in eco-labelling, including acceptance, rejection and unconscious. Findings concur with prior studies about Gen Y consumers' diversity and identify that the expectation can be influenced by brand trust and vice versa. Results also show that consumers' prior shopping experience may affect their acceptance of the eco-labels positively and negatively. Additionally, one of the intriguing findings is that consumers might neglect eco-labels in luxury fashion because they think it is unnecessary.

Finally, the third part of the findings is about communicating eco-labels to consumers, including the design and visibility of eco-labels and their transmissive methods. This part enriches the literature of eco-labels and provides several novel insights. Primarily, based on the design and visibility of eco-labels, this study replenishes the prior research findings. This study suggests that eco-labels could incorporate diverse characteristics, including eco-friendly features, brand iconic attributes, simple design and minor contextual details, to be displayed on one label. Next, this study found that eco-labels could be attached with a brief introduction and visualised on products and packaging. Importantly, these findings cultivate the literature and suggest industrial management to convey eco-labels to consumers.

10.5 Conclusion

This chapter contributes to sustainability communication theory by focusing on insights into eco-labelling strategy in the luxury fashion industry. Findings referred to prior studies, showing a misinterpretation in communicating eco-labels to consumers who may perceive to be confused and misunderstood. A key contribution to the literature shows the diversity of consumer perceptions of eco-labelling, including acceptance, rejection and unconscious. A novelty from the data presents that consumers' prior shopping experience may positively/negatively affect their acceptance of eco-labels. Moreover, another intriguing finding is that consumers may neglect luxury brands' eco-labels because they see them as unnecessary. The next part of the findings is the design and visibility of eco-labels. This study replenishes the literature of eco-labelling in which eco-labels could incorporate diverse characteristics, including eco-friendly features, brand iconic attributes, simple design and minor contextual details, to be displayed on one label. These findings cultivate the literature and suggest industrial management convey eco-labels to consumers.

Therefore, some recommendations are cultivated in managing the eco-labelling strategy for the luxury market. While facing the challenge of low consumer awareness, luxury companies are suggested to increase the presence of eco-labels in the market. To do this, luxury companies can emulate the eco-labelling strategies of the fast fashion or food industries. In terms of eco-labels visualisation and logo design, luxury brands should balance the uniqueness of the luxury industry with its universal implementation on the market. The study thus recommends that luxury companies work on their label design to combine eco-friendly features with the luxury nature of the brand.

This study has several limitations. First, the study draws from a relatively small sample. Thus, it is suggested to investigate this topic through a larger sample to validate the preliminary findings here. Moreover, this study focuses on the luxury fashion industry; the findings of implication may be limited. Future research could imply the findings from this research to explore other luxury sections in different industries. In addition, qualitative research enables a rich insight into consumer perception. Although qualitative research findings are hard to generalise, those provide the basis for future investigation about the topic studied.

References

Achabou, M. A., & Dekhili, S. (2013). Luxury and sustainable development: Is there a match? *Journal of Business Research, 66*(10), 1896–1903.
Amed, I., Balchandani, A., Berg, A., Hedrich, S., Jensen, E. K. & Rölkens, F. (2020, December 1). *The State of Fashion 2021: In search of promise in perilous times.* McKinsey & Company. [Online]. Accessed November 12, 2020, from https://www.mckinsey.com/industries/retail/our-insights/state-of-fashion
Anido Freire, N., & Loussaïef, L. (2018). When advertising highlights the binomial identity values of luxury and CSR principles: The examples of Louis Vuitton and Hermès. *Corporate Social Responsibility and Environmental Management, 25*(4), 565–582.
Arnett, G. (2019). *What the rise of 'ecolabelling' means for retailers.* Vogue Business, 13 December [Online]. Accessed May 13, 2021, from https://www.voguebusiness.com/sustainability/ethical-labelling-selfridges-net-a-porter-kering-allbirds-kering
Athwal, N., Wells, V. K., Carrigan, M., & Henninger, C. E. (2019). Sustainable luxury marketing: A synthesis and research agenda. *International Journal of Management Reviews, 21*(4), 405–426.

Atkinson, L., & Rosenthal, S. (2014). Signalling the green sell: The influence of eco-label source, argument specificity, and product involvement on consumer trust. *Journal of Advertising, 43*(1), 33–45.

Bartiaux, F. (2008). Does environmental information overcome practice compartmentalisation and change consumers' behaviours? *Journal of Cleaner Production, 16*(11), 1170–1180.

Beretti, A., Grolleau, G., & Mzoughi, N. (2009). How cognitive biases can affect the performance of eco-labeling schemes. *Journal of Agricultural & Food Industrial Organization, 7*(2), 1–12.

Bougherara, D., & Combris, P. (2009). Eco-labelled food products: What are consumers paying for? *European Review of Agricultural Economics, 36*(3), 321–341.

Bratt, C., Hallstedt, S., Robèrt, K. H., Broman, G., & Oldmark, J. (2011). Assessment of eco-labelling criteria development from a strategic sustainability perspective. *Journal of Cleaner Production, 19*(14), 1631–1638.

Brécard, D. (2014). Consumer confusion over the profusion of eco-labels: Lessons from a double differentiation model. *Resource and Energy Economics, 37*, 64–84.

Bryman, A. (2016). *Social research methods* (5th ed.). Oxford.

Bucklow, J., Perry, P., & Ritch, E. (2017). The influence of eco-labelling on ethical consumption of organic cotton. In *Sustainability in fashion* (pp. 55–80). Palgrave Macmillan.

Cheah, I., Zainol, Z., & Phau, I. (2016). Conceptualizing country-of-ingredient authenticity of luxury brands. *Journal of Business Research, 69*(12), 5819–5826.

Chen, Y. S., & Chang, C. H. (2013). Greenwash and green trust: The mediation effects of green consumer confusion and green perceived risk. *Journal of Business Ethics, 114*(3), 489–500.

Clemenz, G. (2010). Eco-labeling and horizontal product differentiation. *Environmental and Resource Economics, 45*(4), 481–497.

D'Arpizio, C., Levato, F., Fenili, S., Colacchio, F., & Prete, F., (2020), *How Covid-19 is reshaping the luxury market: The pandemic poses serious challenges to the industry. Here's how to manage through the crisis.* Bain and Company, April 02 [Online]. Accessed May 21, 2021, from https://www.bain.com/insights/how-covid-19-is-reshaping-the-luxury-market-infographic/

D'Arpizio, C., Levato, F., Prete, F., Gault, C., & de Montgolfier, J. (2021, January 14). *The future of luxury: Bouncing back from Covid-19.* Bain & Company [Online]. Accessed June 12, 2021, from https://www.bain.com/insights/the-future-of-luxury-bouncing-back-from-covid-19/

D'Souza, C. (2004). Ecolabel programmes: A stakeholder (consumer) perspective. *Corporate Communications: An International Journal, 9*(3), 179–188.
Davies, I., Oates, C. J., Tynan, C., Carrigan, M., Casey, K., Heath, T., Henninger, C. E., Lichrou, M., McDonagh, P., McDonald, S., McKechnie, S., McLeay, F., O'Malley, L., & Wells, V. (2020). Seeking sustainable futures in marketing and consumer research. *European Journal of Marketing, 54*(11), 2911–2939.
Delmas, M. A., Nairn-Birch, N., & Balzarova, M. (2013). Choosing the right eco-label for your product. *MIT Sloan Management Review, 54*(4), 10–12.
Easterby-Smith, M., Thorpe, R., & Jackson, P. R. (2015). *Management and business research* (5th ed.). Sage.
Eco-Age. (2021). *The GCC Brandmark*. Eco-age [Online]. Accessed June 23, 2021, from http://www.noemptytags.com/archives/eco-age/gcc-brandmark/
Ecolabel Index. (2021). *All ecolabels*. [Online]. Accessed May 21, 2021, from http://www.ecolabelindex.com/ecolabels/
Font, X., Walmsley, A., Cogotti, S., McCombes, L., & Häusler, N. (2012). Corporate social responsibility: The disclosure–performance gap. *Tourism Management, 33*(6), 1544–1553.
Global Ecolabelling Network (GEN). (2021). *What is ecolabelling*. [Online]. Accessed August 23, 2021, from https://www.globalecolabelling.net/what-is-eco-labelling/
Göbbel, T., Goeken, C. & Saadé, H. (2019, June). Decoding Gen Y means recoding your business model. *Roland Berger*. [Online]. Accessed November 6, 2021, from https://www.rolandberger.com/en/About/Events/de/Decoding-Generation-Y/
Gözükara, İ., & Çolakoğlu, N. (2016). A research on generation Y students: Brand innovation, brand trust and brand loyalty. *International Journal of Business Management and Economic Research, 7*(2), 603–611.
Granskog, A., Lee, L., Magnus, K.-H. & Sawers, C. (2020, July 17). *Survey: Consumer sentiment on sustainability in fashion*. McKinsey and Company. [Online]. Accessed December 12, 2020, from https://www.mckinsey.com/industries/retail/our-insights/survey-consumer-sentiment-on-sustainability-in-fashion
Han, J., Seo, Y., & Ko, E. (2017). Staging luxury experiences for understanding sustainable fashion consumption: A balance theory application. *Journal of Business Research, 74*, 162–167.
Henninger, C. E. (2015). Traceability the new eco-label in the slow-fashion industry? Consumer perceptions and micro-organisations responses. *Sustainability, 7*(5), 6011–6032.

ISO. (2019). *Environmental labels*. [Online]. Accessed July 12, 2021, from https://www.iso.org/files/live/sites/isoorg/files/store/en/PUB100323.pdf

Ivan, C. M., Mukta, R., Sudeep, C., & Burak, C. (2016). Long-term sustainable sustainability in luxury. Where else? In *Handbook of sustainable luxury textiles and fashion* (pp. 17–34). Springer.

Janssen, C., Vanhamme, J., Lindgreen, A., & Lefebvre, C. (2014). The Catch-22 of responsible luxury: Effects of luxury product characteristics on consumers' perception of fit with corporate social responsibility. *Journal of Business Ethics, 119*(1), 45–57.

Kearney, M. (2014). The new rules of green marketing: Strategies, tools, and inspiration for sustainable branding. *Journal of Marketing Management, 30*(13–14), 1520–1521.

Kuleto, V., Stanescu, M., Ranković, M., Šević, N. P., Păun, D., & Teodorescu, S. (2021). Extended reality in higher education, a responsible innovation approach for generation y and generation z. *Sustainability, 13*(21), 11814.

Kunz, J., May, S., & Schmidt, H. J. (2020). Sustainable luxury: Current status and perspectives for future research. *Business Research, 13*(2), 541–601.

Malhotra, N. K., Birks, D. F., & Wills, P. (2012). *Marketing research: An applied approach* (4th ed.). Pearson Education.

Osburg, V. S., Strack, M., Conroy, D. M., & Toporowski, W. (2017). Unveiling ethical product features: The importance of an elaborated information presentation. *Journal of Cleaner Production, 162*(20), 1582–1591.

Osburg, V. S., Davies, I., Yoganathan, V., & McLeay, F. (2021). Perspectives, opportunities and tensions in ethical and sustainable luxury: Introduction to the thematic symposium. *Journal of Business Ethics, 169*(2), 201–210.

Pedersen, E. R., & Neergaard, P. (2006). Caveat emptor–let the buyer beware! Environmental labelling and the limitations of "green" consumerism. *Business Strategy and the Environment, 15*(1), 15–29.

Platania, S., Santisi, G., & Morando, M. (2019). Impact of emotion in the choice of eco-luxury brands: The multiple mediation role of the brand love and the brand trust. *Calitatea, 20*(S2), 501–506.

Positive Luxury. (2021). *Brands we've certified*. Positive Luxury [Online]. Accessed June 24, 2021, from https://www.positiveluxury.com/brands/

Rahbar, E., & Wahid, N. A. (2011). Investigation of green marketing tools' effect on consumers' purchase behavior. *Business Strategy Series, 12*(2), 73–83.

Sharma, N. K., & Kushwaha, G. S. (2019). Eco-labels: A tool for green marketing or just a blind mirror for consumers. *Electronic Green Journal, 1*, 42.

Sidiropoulos, E. (2014). Education for sustainability in business education programs: A question of value. *Journal of Cleaner Production, 85,* 472–487.
Taufique, K. M. R., Siwar, C., Talib, B., Sarah, F. H., & Chamhuri, N. (2014). Synthesis of constructs for modeling consumers' understanding and perception of eco-labels. *Sustainability, 6*(4), 2176–2200.
Taufique, K. M. R., Polonsky, M. J., Vocino, A., & Siwar, C. (2019). Measuring consumer understanding and perception of eco-labelling: Item selection and scale validation. *International Journal of Consumer Studies, 43*(3), 298–314.
Testa, F., Iraldo, F., Vaccari, A., & Ferrari, E. (2015). Why eco-labels can be effective marketing tools: Evidence from a study on Italian consumers. *Business Strategy and the Environment, 24*(4), 252–265.
Turunen, L. L. M., & Halme, M. (2021). Communicating actionable sustainability information to consumers: The shades of green instrument for fashion. *Journal of Cleaner Production, 297,* 126605.
Wells, V., Athwal, N., Nervino, E., & Carrigan, M. (2021). How legitimate are the environmental sustainability claims of luxury conglomerates? *Journal of Fashion Marketing and Management: An International Journal, 25*(4), 697–722.

11

'Take a Stand': The Importance of Social Sustainability and Its Effect on Generation Z Consumption of Luxury Fashion Brands

Helen McCormick and Pratibha Ram

11.1 Introduction

Sustainable luxury research has grown exponentially over the past few years; however, focus has been on economic and environmental sustainability, while social sustainability remains under-researched (Kusi-Sarpong et al., 2019). With conscious consumerism on the rise (Kazmi et al., 2021), the socially responsible values of companies are now more than ever in the spotlight. Most social sustainability research focuses on the micro-aspects relating to manufacturing, production such as supply chain, sourcing and labour welfare, while the macro-aspects of social

H. McCormick (✉)
Manchester Fashion Institute, Mancester Metropolitan University, Manchester, UK
e-mail: h.mccormick@mmu.ac.uk

P. Ram
Department of Materials, The University of Manchester, Manchester, UK

© The Author(s), under exclusive license to Springer Nature Switzerland AG 2022
C. E. Henninger, N. K. Athwal (eds.), *Sustainable Luxury*, Palgrave Advances in Luxury, https://doi.org/10.1007/978-3-031-06928-4_11

sustainability such as equality and improving society by responding to the social issues that affect consumers and communities seem to be under-researched. This chapter focuses on luxury brands that are attempting to effect positive social change via their 'sustainability practices' regarding social issues affecting consumers and societies that they operate in (Donato et al., 2020). The 'Great Awokeing' campaign is a call to action, and this has forced brands to respond to and take a stand on societal issues (Deloitte, 2020) regarding diversity, equality and inclusion. Millennial and Generation Z consumers consider social sustainability as an essential principle for their consumption choices, and this segment will account for half of global luxury sales by 2025 (Deloitte, 2020). The purpose of this chapter is to analyse the importance of 'Taking a Stand' to communicate social sustainability regarding societal issues and its effect on Generation Z consumption of luxury fashion brands.

This inquiry focuses exclusively on the luxury fashion sector as the field of application. Real-world evidence is provided by using case studies to show the importance of social sustainability practice in luxury fashion and implications for practice, study limitations and future research directions are discussed. There are no contemporary studies to date that directly consider activism in regard to social sustainability; therefore, it will be addressed in this study utilising Kotler and Sarkar's (2017) social activism as a framework. Empirically, the authors draw upon a novel dataset comprising three global fashion brand case studies.

11.2 Literature Review

Sustainability as an agenda for governments, NGOs, private businesses and consumers worldwide is increasingly paramount to all key stakeholders. *The concept* of *sustainability* is composed of three pillars: Environment, Economic and Social, which have been considered imperative to long-term strategies for achieving sustainable development (WCED, 1987). Often 'Sustainability' is viewed as focusing on environmental quality, natural resource extraction rates and the importance of the global community living within the means of our natural resources (Burns et al., 2019). Due to the fashion industry being noted as the second most pollutant industry globally, there is a rising pressure for fashion brands to communicate what they are doing to become more sustainable. According to Kusi-Sarpong et al. (2019) environmental and economic sustainability is given priority rather than aligning values and beliefs of organisations to benefit society. It is evident that fashion production has a detrimental impact on the environment, and therefore much research has focused on the manufacturing and supply chain aspect; however it could be argued that brands should also consider social sustainability including social equality and social justice (Huang & Rust, 2011) to improve the world we live in.

Social sustainability somewhat overlaps with corporate social responsibility (CSR) as it refers to the well-being and equality of people and society through the management of social resources (Edwards, 2005). Kusi-Sarpong et al. (2019) identified that social sustainability has not gained as much attention within academia; however, it has a major part in creating sustainable communities and relationships between brands and consumers. The UN (2022) focus on the '*social dimension of corporate sustainability, of which human rights is the cornerstone*'. They go on to state that businesses' social licence 'to operate depends on their social sustainability efforts', which include inequality, gender equality, women's empowerment, people with disabilities, poverty, education and health as a few of the issues to address.

The global pandemic has been a 'wake-up call' regarding sustainability, with consumers increasingly considering what they are consuming and the impact their consumption behaviour has on the environment (Raconteur, 2020). Schaefer (2020) identified a preference for purpose-driven brands, rather than simply for profit. There is a rise in brands starting to hold themselves accountable regarding environment and social issues and there has been movement for brands to publish how they are a social sustainable brand and what they stand for in society. Reimann et al. (2012) suggest that consumers do not just buy the product but buy the brand, rather buying into the purpose of the brand. Purpose is the core principle of a company and not cause marketing; purpose-driven brand communications is about brands taking action (Neff, 2019). Sustainable marketing orientation of a purpose-driven brand is achieved by integrating social, environmental and economic sustainability, and proactive development of strategies that would benefit consumers and the society through ongoing ethical marketing initiatives (Lučić, 2020).

Porter and Kramer's (2019) research concerning creating shared value identified that companies must reframe their role in society, moving away from a capitalist system to consider the people aspect of business strategy.

The luxury fashion market is significant and has continued to grow over the past 10 years, reaching 45 billion US dollars in 2020 (Farfetch, 2019), still growing by 8.2% from 2020 to 2021 during the Covid-19 pandemic (Deloitte, 2020). Due to aspects associated to luxury goods and the longevity associated to luxury brands, they are more aligned to sustainable practices than fast fashion brands, for example, that focus on trend-led/disposable fashion. The perception of what defines a luxury brand has evolved over the years including aspects such as being artistic, craftmanship, international presence (Chevalier & Mazzalovo, 2012), service (Kapferer & Bastien, 2008), quality and high price (Dubois et al., 2001). Ko et al. (2019) most recently concluded that the perception of a luxury brand is dependent on consumers' evaluation of quality, authenticity, prestigious image, premium price and resonance. The final aspect resonance refers to Keller et al.'s (2011) brand resonance model, referring to the relationship between brand and consumer, leading to behavioural loyalty, attitudinal attachment, sense of community and active engagement. In today's polarised and global fashion marketplace, resonance to values and purpose is becoming increasingly important. It is inherent that a luxury brand strategy aligns with the current and target customers' values; if consumers cannot identify with the brand, they will not consume it.

According to Highsnobiety, it is forecast that millennials and Gen Z will make up more than 60% of the luxury market by 2026, adding 500 billion US dollars to luxury sales (Morency, 2021). Currently Gen Z is responsible for 32% of sales in the luxury goods market, setting to increase to 45% by 2025 (Farfetch, 2019). A survey conducted by First Insight identified that Generation Z are most influenced and invested in fashion brands that have sustainable business practices. One of the reasons that this generation holds more sustainable values could be that they are being educated about climate change and are much more aware of environmental, political and social problems than previous generations (UN, 2022). Due to Gen Z's awareness, there is an expectation that the brands that they consume from show respect to the planet, cultures and society (Rambourg, 2020). According to McKinsey (2022), nine out of ten believe that it is the brands/businesses that have a responsibility to address environmental and social issues. In fact, it was reported by A T Kearney (2020) that half of 15–20-year-olds would boycott brands if they did not agree with its ethics, self-identifying themselves as 'woke' alert to injustice in society (Sobande, 2019; McColl et al., 2021), therefore it is important that the communication concerning the sustainable initiatives undertaken by a company are part of its strategic marketing strategy. This generation values brands that support meaningful topics such as climate change and human rights; therefore, brands now need to share information about not only their practices but socially responsible values.

According to research conducted by McColl et al. (2021) which considered social campaigns, the most successful were brands where the cause was part of the brand DNA, meaning it was part of the brand purpose, which allowed the consumer to connect and appealing to ethical ideals. They further identified that the cause must identify and align with the target consumer value; otherwise fake authenticity or 'woke-washing' without any intention of making changes can be detrimental to the brand. Therefore, it is paramount that brands identify their purpose and values to communicate their stance to connect with consumers, particularly if they are trying to attract Gen Z who select brands based on their sense of social responsibility (Rodriguez et al., 2019). Ritch and McColl (2021) discuss 'Woke branding', building on a brand's purpose to be socially responsive in relation to sustainability ideology. Fashion businesses have moved away from push-marketing towards creating connection through storytelling and engagement. They also go on to state that marketers need to construct authentic campaigns that 'nudge' customers towards sustainable practices, fit the purpose of the brand and are authentic, aligning sustainable principles of both consumer and the brand (Ritch & McColl, 2021).

It is evident that brands can no longer just market products based on performance characteristics; this is why we have seen a rise over the past few years in what Kotler and Sarkar (2017) define as 'Social Activism'. Brands have a values-driven agenda to support society and the planet and stress the importance of justice and fairness for all people (Kotler &

Sarkar, 2017). There is a complexity concerning contemporary activism as brands are communicating a message for change which can include or exclude people; however, Vogue Business proclaims that *'Brands used to avoid political messaging for fear of offending clients with different views. The danger now is not being committed enough'*. Over recent years specific social activist issues that brands have been championing are movements such as Black Lives Matter, which gained global traction after the death of George Floyd in America, highlighting the murder of an innocent man by an American policeman (Lebron, 2017). Brands showed support of the Black Lives Matter protest using Instagram to launch Blackout Tuesday, posting a black square with #BlackoutTuesday. Many luxury brands including Nordstrom, Gucci, Alexander McQueen and Prada posted to show support and solidarity with some brands demonstrating support with significant donations to organisations and initiatives combating systemic racism, social justice and education. A study by Shendruck and Bain (2021) several years after the event considered behaviours of fashion brands in regard to representation within campaigns; however, they found limited change concerning diversity and inclusivity of Black models.

Brands are more proactive concerning expressing their support to communities such as LGBTQ+, evident by the 25% increase in merchandise and products during Pride in 2019 (Chitrakorn, 2020) and events such as International Women's Day to build equal rights for *women* and girls worldwide with communication from fashion brands increasing 45% according to Edited (2019). Gen Z are interested in brands that have moral and ethical qualities, seeking brands with passion and that respond and react to the global environment. It is imperative that brands communicate what it is that they are doing to support social sustainability through their platforms.

11.3 Conceptual Framework

11.3.1 Social Dimension in Brand Activism

Brand activism is driven by sense of justice and fairness for all. Sarkar and Kotler (2018, p. 554) define Brand activism as '*business efforts to promote, impede or direct social, political, economic, and/or environmental reforms with the desire to promote or impede improvements in society*'. They further divide brand activism into six categories—social, cultural, political, legal, environmental and economic. Brands embrace a few or a combination of these depending on their priorities and as basis of their brand activism (McColl et al., 2021). The choice of brand activism also depends on the ethical choices of the consumers that the brand is targeted towards (Guèvremont, 2017).

Social brand activism has increased in popularity; many brands seem to be supporting and even taking a stand against social and political injustice by integrating these issues within either the sustainability policies and practices or their branding and marketing strategies (Amed et al., 2019; Bakhtiari, 2021). In recent times brands signal their responsibility towards socio-political issues by supporting movements such as Me Too, Black Lives Matter, Say Her Name, Pro Choice and many more. Many of these socio-political issues are controversial and divisive in nature which means there would be consumers and other stakeholders that are supportive while others may be against (Vredenburg et al., 2020). Social brand activism is defined as a brand's public demonstration by way of statements and/or actions of either support for or opposition to one side of the socio-political issue (Bhagwat et al., 2020, p. 1).

11.3.2 From Social Sustainability to Social Activism

The social dimension of sustainability has grown in prominence and is widely considered as important; however it is still under-researched and does not have a single or widely contested definition (Tiainen, 2016; Woodcraft, 2012). According to Sachs (1999) social sustainability rests on effective distribution of fair human rights within political, social, economic and culture dimensions. The institution perspective according to Littig and Griessler (2005, p. 72) is *'to work within society and related institutions that satisfy a set of human needs in a way that fulfils social justice, equity and participation'*. Similarly, sense of justice and fairness for all are some of the commonalities between social sustainability and social activism. However, they also differ in many ways and it's important to discuss these differences to gain a nuanced ideation of social activism concerning what it is and what it is not.

Social sustainability is built on specific goals, action plan, results and mechanism to measure these results (Wettstein & Baur, 2016). Social activism on the other hand is meant to increase awareness and to support social and/or behavioural change (Vredenburg et al., 2020), and there is no fixed response or mechanism to address social and political issues (Korschun et al., 2019). Compared to social sustainability, activism may not always be linked to brands' core business focus (Dodd & Supa, 2014). One of the most crucial differences is the nature of the social cause or issue: the more divisive and controversial the social issue, the more likely it would be categorised as social activism. Social sustainability is usually concerned with prosocial causes that are well accepted for instance supporting child education (Mukherjee & Althuizen, 2020). On the other hand, causes such as Prochoice, Trans athletes, Taking the Knee tend to be divisive, controversial and generate a passionate response. Where there is a heated debate that elicits polarising opinions within consumer and stakeholder response, the brand needs to 'Take a Stand' and choose a specific social or political ideology related to the cause (Hydock et al., 2020). Such an act of 'Taking a Stand' has the potential to dissatisfy certain consumer segments and the brand could risk losing sales and profit (Park & Jiang, 2020). It is due to this risk that social activism is not directly linked to brands' core business focus, in comparison to social sustainability (Dodd & Supa, 2014).

11.3.3 A Nested Approach to Social Activism

For the purpose of this chapter, drawing from the above literature review, we propose the following conception and dimensions of social activism.

While there are overarching themes between social sustainability and social activism, they are not always mutually dependent or the same. We do not delineate these concepts but suggest that the nature and the type of social cause must be considered to categorise if a specific sustainability policy, practice or procedure is either related to social sustainability or social activism. This conception underlines one fundamental dimension, supporting prosocial causes that are considered well accepted charitable donations for health for instance is social sustainability, whereas supporting controversial and divisive causes is social activism. However, the interpretation of 'Contested, controversial, and polarising' cause may change over time and depends on the culture/country it is applied; therefore, it is contextual and interpretive (Vredenburg et al., 2020; Nalick et al., 2016). An example being supporting social causes such as women's empowerment and accepting LGBTQ+ community in many countries and cultures; however there are certain cultures and countries where these causes are still contested and controversial, and polarising and supporting these causes will be considered social activism in those cultures and countries. We define social brand activism as '*Taking a stand against contested, controversial, and polarising social issues by raising awareness and by taking actions to fight social injustice*'.

The following sections discuss this ideation of social activism nested within social sustainability (see Fig. 11.1) with evidence of real-world business cases.

11.4 Analysis of Case Studies

A comparative case study approach (Eisenhardt, 1989; Holt, 2004; Miles & Huberman, 1994; Yin, 2003), was used to analyse the social sustainability and social activism initiatives reported by the luxury brands.

11.4.1 Case Study Sample

Luxury brands that are considered the industry leaders and those that have received distinction for corporate sustainability by The Standard

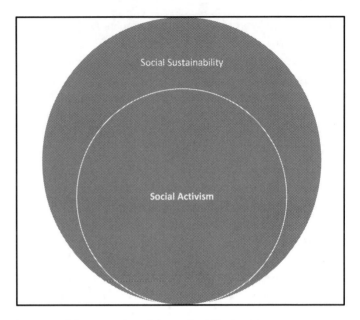

Fig. 11.1 Conceptual framework social sustainability/activism

and Poor's (S & P) Global Corporate Sustainability Assessment Index (CSA) for the year 2021 were chosen. The reasoning for the case study sampling selection criteria is solely because the S & P Global Corporate Sustainability Assessment Index (CSA) associated with Dow Jones Sustainability Index has been the basis for various Environmental, Social, and Governance (ESG) indices for the last two decades. It is a trusted source and is and a reference point for corporate sustainability practices within all global businesses including retail and fashion industry.

All the luxury brands under the category of Textiles, Apparel & Luxury Goods in the S & P Corporate Sustainability Assessment Index, 2021 were chosen as samples. The sustainability practices and initiatives that these brands report on their corporate websites were analysed. Corporate websites of brands were chosen for case study analyses because this is often a standard platform utilised globally, it is a point of purchase channel and brand engagement for consumers and finally corporate websites are often the reference point for investors. A representative selection of three cases from the total sample analysed is presented below,

11.4.2 Case 1: Moncler S.p.A.

Moncler received the highest distinction of S & P Global Gold Class in 2021 for the third year in a row. It is the industry leader for the category of Textiles, Apparel & Luxury Goods in the S & P World and Europe indices (S & P, 2022). '*The increasing integration between business decisions and the assessment of their environmental and social impacts is at the core of the Group's ability to create long-term value for all its stakeholders*' (Moncler, 2022). Its Strategic Sustainability Plan is referred to as 'Born to Protect'—protecting the planet and people. The sustainable practices as reported by Moncler are categorised as following:

- **Responsible Business Management**—This is primarily concerned with the code of ethics in addition to corporate and business operation of Moncler.
- **Think Circular**—Product development, packaging, quality and safety.
- **Act on Climate Change**—Environmental policy and management.
- **Nurture Genius**—Employee and industrial practices, such as diversity, equal opportunities, occupational health, safety and well-being.
- **Be Fair**—Initiatives that relate to supply chain, sourcing, production, counterfeits, understanding and fulfilling customer expectations.
- **Give Back**—Supporting growth of local communities, by supporting scientific research, Covid relief, refugees.

Kusi-Sarpong et al. (2019) observed that many brands give priority to environmental and economic sustainability over social sustainability; this disparity can also be observed with Moncler: the brand is oriented more towards environmental sustainability than social sustainability. Social sustainability is limited to practices related to business, employee, supply chain and supporting local communities. The corporate website does not have any indication of Moncler's social activism. Moreover, drawing from the ideation of social activism discussed in the sections above practices such as equal opportunities employer, safety and well-being Covid relief and refugee aids cannot be classified as social activism. Initiatives such as equal opportunities employer, safety and well-being would be considered standard business ethics, while Covid relief and refugee aids are well-accepted pro-social causes; these are socially sustainable but not social activism.

11.4.3 Case 2: Burberry Group plc

Burberry received the highest distinction of S & P Global Gold Class in 2021 indices (S & P, 2022). *'As a purposeful, values-driven brand, we are committed to being a force for good in the world'* (Burberry, 2022). The sustainable practices as reported by Burberry are categorised as following:

- **Environment**—Initiatives include balancing emissions, promoting renewables, transparency of the supply chain, use of sustainable materials, ethical sourcing and minimising waste.
- **Communities**—Fair employment practices; some of the initiatives are aimed at youth and disadvantaged communities, supporting literacy and efforts to combat the Covid pandemic.
- **People**—This category focuses on initiatives for diversity and inclusion that focuses on causes such as:
    1. **Women empowerment**—#BurberryVoices initiative event was organised on The International Women's Day and donations were given for charities such as London Youth and The Prince's Trust Women.
    2. **Supporting LGBTQ+ community**—Donations given to charity organisations such as UK Black Pride, Albert Kennedy Trust and Stonewall.

On the outset grouping the initiatives under environment, community and people may seem like Burberry gives equal importance for both environmental and social sustainability, but this can only be established by analysing the value and the type of resources allocated for the different categories and the specific cause. Initiatives aimed at the youth and the disadvantaged communities such as supporting literacy and efforts to combat the Covid pandemic are social sustainability practices. Burberry supports causes such as women's empowerment and LGBTQ+

community by donating to charities in the UK. One may argue that these causes are well-accepted pro-social causes and are not 'Contested, controversial, and polarising' (Vredenburg et al., 2020; Nalick et al., 2016) in the UK context. Supporting these causes are possibly considered as standard practice among majority of other organisations within the UK, in which case it may not be classified as social activism. The causes of women empowerment and supporting LGBTQ+ community are critical, and this has now been included in the code of ethics and corporate governance by many organisations. The sustainability practices reported in Burberry's corporate website suggest that the brand has aligned social sustainability practices; however, drawing on the conception of social activism from the above sections, within the UK these initiatives may not fall under the category of a social activism.

11.4.4 Case 3: Gucci (Kerring Group)

The group received S & P Silver Gold Class in 2021 indices (S & P, 2022), *'Far more than an ethical necessity, sustainability is a driver of innovation and value creation for the Group, its Houses, and its stakeholders'* (Kerring, 2022). The sustainable practices as reported by Gucci are categorised as following:

* **Planet**—Sustainable initiatives are focused on environment, circular economy, green energy and animal welfare.
* **People**—Supporting people and community worldwide such as volunteering, refugees, Covid vaccines delivery, scholarships. In addition, it supports social and societal issues such as

    1. **Gender equality**—Showcase women's contribution to the progress of humanity, feminist movements and leadership action, menstrual stigma, intersectional feminism, female genital mutilation (FGM), abuse epidemic, cyber harassment, obstetric violence, sexual harassment, victims of acid attacks, child marriage and raising awareness on the issue of domestic violence.

2. **Diversity and inclusion**—Support human rights, LGBTQ+ community, inclusion of people with disabilities, diversity, immigration and inclusion.
3. **Social Justice**—Racial justice, human trafficking and modern slavery, migrants and refugees' experiences and equitable representation, intimate partner violence, Black history month, refugees, child marriage, Prochoice campaigns, The #SayHerName campaign brings awareness to the often-invisible names and stories of Black women and girls victimised by police violence transcription and provides support to their families.

In comparison to Moncler and Burberry, Gucci (Kerring) seems to balance both environmental and social sustainability practices. It has integrated both social sustainability and social activism practices. Supporting causes such as human trafficking and modern slavery, cyber harassment, domestic violence and inclusion of people with disabilities may be considered social sustainability as they are universally accepted responses for these issues (Korschun et al., 2019; Nalick et al., 2016). However, issues such as menstrual stigma, intersectional feminism, female genital mutilation (FGM), obstetric violence, victims of acid attacks, child marriage, immigration and refugees, Prochoice, Black women and girls victimised by police violence are highly contested, controversial and polarising as there are consumers and other stakeholders that may fall on either side of the debate and others that are not even aware of these social issues (Vredenburg et al., 2020). By supporting these issues Gucci seems to follow a social activist approach as advocated by Sarkar and Kotler (2018, p. 554). Gucci has taken a stand, supported and made revelations that may be shocking to some by demonstrating its support to polarising issues. This could be considered controversial, but this also demonstrates exemplarity social activism (Han et al., 2017). It signals social consciousness, strives to not just bring awareness to social issues but also takes a stand, even at the expense of losing certain consumers and stakeholders (Mukherjee & Althuizen, 2020). Such activist brands aim to bring awareness to controversial social issues which may even reform the moral judgement of the general masses (Sibai et al., 2021).

11.5 Implications for Managers: Why Is Social Activism Important?

This chapter presented three business cases of luxury brands that are ranked top in the world for corporate sustainability by the S & P sustainability index, however only Gucci (Kerring group) can be considered a social activist. The other two business cases of Moncler and Burberry even though they were awarded the distinction of Gold class by the S & P index, they seem to prioritise environmental initiatives over social sustainability. While Burberry does practise social sustainability by supporting women empowerment and LGBTQ+ community the initiatives cannot be considered social activism.

In addition to supporting social issues Gucci has also contributed to foster a paradigm shift in the luxury business model by supporting and practising social activism which may enhance awareness of consumers to be more socially aware and socially sustainable (Donato et al., 2020). Such social activism also builds a strong consumer–brand bond specifically with consumers that care about and expect the brand to take a stand (Koch, 2020; Schmidt et al., 2021). The necessity to embrace pro-social actions also brings in an economic opportunity, business growth and a competitive advantage (Fisk, 2010).

Gucci (Kerring) has been successful in integrating social activism within their social sustainability framework, additionally it seems to give somewhat equal importance for both environmental and social sustainability. This trend must be embraced by other luxury brands; social activism must be viewed as a value-driven opportunity for a purpose-driven engagement with the target consumers and for brand differentiation (Tata et al., 2013). Social sustainability must be an integral part of its brand's purpose and not an afterthought, and before choosing a specific social issue, the brand must ensure that it passionately believes in the cause and the cause aligns with the ideals of their consumers and the chosen issue is important to them (Kotler & Sarkar, Kotler & Sarkar, 2017; Sarkar & Kotler, 2018, p. 554). If not, it may be perceived as 'woke-washing' and disingenuous (Watson, 2019). This is specifically important for the Millennials and the Gen Z, who are more sustainable, conscious, expect their brands to take action and are willing to boycott brands that are not considered sustainable (Amed et al., 2019).

11.6 Concluding Remarks

Consumers are more conscious of the environmental and social implications of their fashion choices and counteract this issue by demanding brands to make changes (Henninger et al., 2016). With the rise in woke consumerism especially among the millennials and Gen Z, they require the brands to also share their commitment by signalling their support for fighting social injustice by 'Taking a Stand' (Hess, 2016; Hydock et al., 2020). Luxury brands need to catch up with a need for great urgency, millennials believe luxury brands need to do 'more good' and not just 'less bad' (McColl et al., 2021). Social responsibility and social sustainability are no longer a preference, and it is now an absolute requirement; they are as important as environmental issues and not just something nice-to-have (Winston, 2016).

References

Amed, I., Balchandani, A., Beltrami, M., Berg, A., Hedrich, S., & Rölkens, F. (2019, February). *The influence of 'woke' consumers on fashion*. McKinsey & Company. Accessed February 02, 2022, from https://www.Mckinsey.com/industries/retail/our-insights/the-influence-of-woke-consumers-on-fashion

AT Kearney. (2020). *Sustainability report*. Accessed February 20, 2022, from https://www.kearney.com/documents/20152/107282429/Kearney+Sustainability+Report+2020.pdf/5811a7c9-a5f7-162b-59fb-ada1701e8786?t=1618593648000

Bakhtiari, K., (2021). Climate emergency, gen-z voices and brand activism. *Forbes*. Accessed August 20, 2021, from https://www.forbes.comsiteskianbakhtiari20210730how-can-sustainable-brands-win-over-gen-z-activistssh73fa83f52115

Bhagwat, Y., Warren, N. L., Beck, J. T., & Watson, G. F., IV. (2020). Corporate socio-political activism and firm value. *Journal of Marketing, 84*(5), 1–21.

Burberry Approach to Responsibility. (2022). *Burberry Inc.* Accessed February 12, 2022, from https://www.burberryplc.com/en/responsibility/approach-to-responsibility.html

Burns, H. L., Kelley, S. S., & Spalding, H. E. (2019). Teaching sustainability: Recommendations for best pedagogical practices. *Journal of Sustainability Education, 19.*

Chevalier, M., & Mazzalovo, G. (2012). *Luxury brand management: A world of privilege* (2nd ed.). Wiley.

Chitrakorn, K. (2020). *How fashion got marketing right in 2020.* Accessed December 15, 2021, from https://www.voguebusiness.com/companies/how-fashion-got-2020-marketing-right-politics-tiktok-gaming-fashion-weeks

Deloitte Global. (2020). *The global power of luxury goods: The new age of fashion and luxury.* Accessed February 02, 2022, from https://www2.deloitte.com/global/en/pages/consumer-business/articles/gx-cb-global-powers-of-luxury-goods.html

Dodd, M. D., & Supa, D. W. (2014). Conceptualizing and measuring "corporate social advocacy" communication: Examining the impact on corporate financial performance. *Public Relations Journal, 8*(3), 2–23.

Donato, C., Buonomo, A., & Angelis, M. (2020). Environmental and social sustainability in fashion: A case study analysis of luxury and mass-market brands. In S. S. Muthu & S. S. Gardetti (Eds.), *Sustainability in the textile and apparel industries: Consumerism and fashion sustainability* (pp. 71–85). Springer Nature.

Dubois, B., Laurent, G., & Czellar, S. (2001). *Consumer rapport to luxury: Analyzing complex and ambivalent attitudes* (Vol. 736). HEC Paris.

Edwards, A. R. (2005). *The sustainability revolution: Portrait of a paradigm shift.* New Society Publishers.

Eisenhardt, K. M. (1989). Building theories from case study research. *Academy of Management Review, 14*(4), 532–550.

Farfetch. (November 20, 2019). Generation Y and Z share of global personal luxury good sales in 2017 and 2025 [Graph]. In *Statista.* Retrieved June 23, 2022, from https://www.statista.com/statistics/1092048/gen-y-and-z-share-of-global-personalluxury-good-sales/

Fisk, P. (2010). *People planet profit: How to embrace sustainability for innovation and business growth.* Kogan Page.

Gucci Equilibrium Sustainability Strategy. (2022). *Gucci (Kerring Group)*. Accessed February 12, 2022, from https://equilibrium.gucci.com/gucci-sustainability-strategy/

Guèvremont, V. (2017). The Convention in other international forums: A crucial commitment. In *Reshaping cultural policies: A decade promoting the diversity of cultural expressions for development. 2005 Convention global report* (pp. 143–161). UNESCO.

Han, J., Seo, Y., & Ko, E. (2017). Staging luxury experiences for understanding sustainable fashion consumption: A balance theory application. *Journal of Business Research, 74*, 162–167.

Henninger, C. E., Alevizou, P. J., & Oates, C. J. (2016). What is sustainable fashion? *Journal of Fashion Marketing and Management: An International Journal, 20*.

Hess, A. (2016). Earning the 'woke' badge. *The New York Times Magazine*. Accessed February 12, 2022, from: https://www.nytimes.com/2016/04/24/magazine/earning-the-woke-badge.html

Holt, D. B. (2004). *How brands become icons: The principles of cultural branding*. Harvard Business Press.

Huang, M. H., & Rust, R. T. (2011). Sustainability and consumption. *Journal of the Academy of Marketing Science, 39*(1), 40–54.

Hydock, C., Paharia, N., & Blair, S. (2020). Should your brand pick a side? How market share determines the impact of corporate political advocacy. *Journal of Marketing Research, 57*, 1135–1151.

Kapferer, J. N., & Bastien, V. (2008). Luxe oblige (No. hal-00786816).

Kazmi, S. H. A., Shahbaz, M. S., Mubarik, M. S., & Ahmed, J. (2021). Switching behaviors toward green brands: Evidence from emerging economy. *Environment, Development and Sustainability, 2021*, 1–25.

Keller, K. L., Parameswaran, M. G., & Jacob, I. (2011). *Strategic brand management: Building, measuring, and managing brand equity*. Pearson Education India.

Ko, E., Costello, J. P., & Taylor, C. R. (2019). What is a luxury brand? A new definition and review of the literature. *Journal of Business Research, 99*, 405–413.

Koch, C. H. (2020). Brands as activists: The Oatly case. *Journal of Brand Management, 27*(5), 593–606.

Korschun, D., Rafieian, H., Aggarwal, A., & Swain, S.D., (2019). *Taking a stand: Consumer responses when companies get (or don't get) political*. Accessed from SSRN 2806476.

Kotler, P., & Sarkar, C. (2017). Finally, brand activism. *The Marketing Journal, 9*, 2017.
Kusi-Sarpong, S., Gupta, H., & Sarkis, J. (2019). A supply chain sustainability innovation framework and evaluation methodology. *International Journal of Production Research, 57*(7), 1990–2008.
Lebron, C. J. (2017). *The making of black lives matter: A brief history of an idea.* Oxford University Press.
Littig, B., & Griessler, E. (2005). Social sustainability: A catchword between political pragmatism and social theory. *International Journal of Sustainable Development, 8*(1–2), 65–79.
Lučić, A. (2020). Measuring sustainable marketing orientation—Scale development process. *Sustainability, 12*(5), 1734.
McColl, J., Ritch, E. L., & Hamilton, J. (2021). *Brand purpose and 'woke' branding campaigns. New perspectives on critical marketing and consumer society.* Emerald Publishing Limited.
McKinsey. (2022). *State of Fashion 2022: Uneven recovery and new Frontiers.* McKinsey and Company. Accessed August 20, 2021, from https://www.mckinsey.com/industries/retail/our-insights/state-of-fashion
Miles, M. B., & Huberman, A. M. (1994). *Qualitative data analysis: An expanded sourcebook.* Sage.
Moncler Group. (2022). *Consolidated non-financial statement.* Accessed January 02, 2022, from https://www.monclergroup.com/sustainability/Moncler-Consolidated-non-financial-statement-2020-DEF/data/Moncler%20-%20Consolidated%20non%20financial%20statement%202020.pdf
Morency, C. (2021). *Why the future of the luxury market will be dictated by youth culture, Highsnobiety.* Accessed February 02, 2022, from What Will Matter Most to the New Luxury Consumer? | Highsnobiety.
Mukherjee, S., & Althuizen, N. (2020). Brand activism: Does courting controversy help or hurt a brand? *International Journal of Research in Marketing, 37*(4), 772–788.
Nalick, M., Josefy, M., Zardkoohi, A., & Bierman, L. (2016). Corporate sociopolitical involvement: A reflection of whose preferences? *Academy of Management Perspectives, 30*(4), 384–403.
Neff, J. (2019). *Purpose isn't cause marketing—How to know the difference.* Accessed December 20, 2021, from https://adage.com/article/cmo-strategy/purpose-isnt-cause-marketing-how-know-difference/2179321.

Park, K., & Jiang, H. (2020). Signaling, verification, and identification: The way corporate social advocacy generates brand loyalty on social media. *International Journal of Business Communication*, p. 2329488420907121.

Porter, M. E., & Kramer, M. R. (2019). Creating shared value. In *Managing sustainable business* (pp. 323–346). Springer.

Raconteur. (2020). *The future customer*. Accessed January 02, 2022, from future-customer-2020__1_.pdf (ctfassets.net)

Rambourg, E. (2020). *Future luxe: What's ahead for the business of luxury*. Publishing Group West.

Reimann, M., Castaño, R., Zaichkowsky, J. L., & Bechara, A. (2012). How we relate to brands: Psychological and neurophysiological insights into consumer-brand relationships. *Journal of Consumer Psychology, 22*(1), 128–142.

Ritch, E. L., & McColl, J. (Eds.). (2021). *New perspectives on critical marketing and consumer society*. Emerald Publishing.

Rodriguez, M., Boyer, S., Fleming, D., & Cohen, S. (2019). Managing the next generation of sales, gen Z/millennial cusp: An exploration of grit, entrepreneurship, and loyalty. *Journal of Business-to-Business Marketing, 26*(1), 43–55.

S & P. (2022). The sustainability yearbook - 2022 rankings. *The standard and Poor's global*. Accessed January 12, 2022, from https://www.spglobal.com/esg/csa/yearbook/2022/ranking/

Sachs, I. (1999). Social sustainability and whole development: Exploring the dimensions of sustainable development. In B. Egon & J. Thomas (Eds.), *Sustainability and the social sciences: A cross-disciplinary approach to integrating environmental considerations into theoretical reorientation*. Zed Books.

Sarkar, C., & Kotler, P. (2018). *Brand activism: From purpose to action*. Idea Bite Press. Accessed from http://www.activistbrands.com/the-book/

Schaefer, M. (2020). Marketing rebellion: The most human company wins. *Journal of Applied Communications, 104*(3), 1–4.

Schmidt, H. J., Ind, N., Guzman, F., & Kennedy, E. (2021). Sociopolitical activist brands. *Journal of Product and Brand Management*.

Shendruck, A., & Bain, M. (2021). *An analysis of 27,000 Instagram images show that fashion's BLM reckoning was mostly bluster*. Accessed January 02, 2022, from Fashion brands aren't keeping their Instagram diversity promises — Quartz (qz.com)

Sibai, O., Mimoun, L., & Boukis, A. (2021). Authenticating brand activism: Negotiating the boundaries of free speech to make a change. *Psychology & Marketing, 38*(10), 1651–1669.

Sobande, F. (2019). Woke-washing: "Intersectional" femvertising and branding "woke" bravery. *European Journal of Marketing, 54*(11), 2723–2745.

Tata, R., Stuart, H., & Sarkar, C., (2013). Why making money is not enough. *MIT Sloan Management Review*. Accessed January 12, 2022, from https://sloanreview.mit.edu/article/why-making-money-is-not-enough/

Tiainen, H. (2016). Contemplating governance for social sustainability in mining in Greenland. *Resources Policy, 49*, 282–289.

UN. (2022). Accessed January 12, 2022, from Social Sustainability |UN Global Compact.

Vredenburg, J., Kapitan, S., Spry, A., & Kemper, J. A. (2020). Brands taking a stand: Authentic brand activism or woke washing? *Journal of Public Policy & Marketing, 39*(4), 444–460.

Watson, I. (2019). *Unilever CEO Alan Jope laments the 'woke-washing' ads 'polluting' brand purpose*. Accessed December 15, 2021, from https://www.thedrum.com/news/2019/06/19/unilever

WCED, S. W. S. (1987). World commission on environment and development. *Our Common Future, 17*(1), 1–91.

Wettstein, F., & Baur, D. (2016). Why should we care about marriage equality? Political advocacy as a part of corporate responsibility. *Journal of Business Ethics, 138*(2), 199–213.

Winston, A. (2016). Luxury brands can no longer ignore sustainability. *Harvard Business Review, 8*(2), 1–3.

Woodcraft, S. (2012). Social sustainability and new communities: Moving from concept to practice in the UK. *Procedia-Social and Behavioral Sciences, 68*, 29–42.

Yin, R. K. (2003). Designing case studies. *Qualitative Research Methods, 5*(14), 359–386.

12

Chinese Consumer Attitudes Towards Second-Hand Luxury Fashion and How Social Media eWoM Affects Decision-Making

Rosy Boardman, Yuping Zhou, and Yunshi Guo

12.1 Introduction

The second-hand luxury fashion market has grown rapidly in recent times and is forecast to double over the next 5 years to reach $77 billion by 2025 (Diderich & Theodosi, 2021). The rapid growth of sales channels such as peer-to-peer marketplaces, apps dedicated to second-hand sales and specialised online platforms has given second-hand luxury fashion items global prominence (Turunen et al., 2020). Thus, digital channels and the second-hand luxury fashion industry now go hand in hand. Social media is one of the primary sources of fashion inspiration for consumers due to the fact that garments and accessories are displayed on real people in everyday settings, making it easier for people to see how items could look on themselves (Chrimes & Boardman, 2022). Although there is a plethora of social media research, the majority of studies focus on

R. Boardman (✉) • Y. Zhou • Y. Guo
Department of Materials, The University of Manchester, Manchester, UK
e-mail: Rosy.Boardman@manchester.ac.uk

Western social media platforms, with more research focusing on Chinese social media platforms needed, as findings across cultures and regions may differ (Yao et al., 2019). Moreover, while scholars have studied the influence of electronic word-of-mouth (eWoM) on consumer attitudes, this has not been explored in the context of second-hand luxury fashion, particularly in the context of the Chinese market. Chinese consumers are the biggest consumers of luxury items in the world, with 45% of the market share (Bain and Company, 2021), and this appetite for luxury fashion is now translating into the second-hand luxury market in China. Yet, it is unknown how Chinese consumers feel about second-hand luxury fashion and whether there are any unique challenges that this industry may face in this country context. With the continuous expansion of China's second-hand luxury fashion market and the increasing popularity of social media platforms, exploring how eWoM affects Chinese consumers' decision-making towards second-hand luxury fashion will provide interesting insights for academics and practitioners alike. As such, this chapter fills a gap in the literature by exploring Chinese consumers' attitudes towards second-hand luxury fashion and how eWoM on social media affects their decision-making towards second-hand luxury fashion purchases. This aim will be achieved by answering the following Research Questions:

RQ1: What are the Attitudes of Chinese consumers towards second-hand luxury fashion?
RQ2: How does eWoM on social media affect Chinese consumers' decision-making in relation to second-hand luxury fashion?

12.2 Literature Review

12.2.1 Second-Hand Luxury Fashion and the Chinese Market

Sustainable consumption focuses on better pacing of behaviour in a way that limits the depletion of the Earth's natural resources whilst also reducing the use of toxins that may put those resources in jeopardy for future generations (Banbury et al., 2012). Usually, second-hand products are bought and sold again multiple times by different people (Machado et al., 2019), thereby prolonging the life of items and encouraging a more sustainable method of consumption. Roux and Guiot (2008) define second-hand consumption as the acquisition of used items through specific means and places of exchange. Hence, the second-hand consumption business model, where customers are also the main partners and suppliers, helps reduce resource use and waste (Gopalakrishnan & Matthews, 2018).

Exclusivity is considered a distinctive characteristic of luxury (Fionda & Moore, 2009), which is maintained through limited production. Luxury is associated with craftsmanship and respect for heritage (Zhang & Cude, 2018). Therefore, luxury fashion products are described as exceptional, sophisticated and exquisite goods (Chevalier & Mazzalovo, 2012). Besides their functional value, Gao et al. (2009) purport that luxury fashion goods, including clothing and accessories, convey a specific brand image, providing the owner with a certain status.

Second-hand goods are conceptualised as those that have been previously owned and used (Carrigan et al., 2013). Second-hand luxury products still offer the exclusivity, craftsmanship, heritage and status as luxury products due to their brand name and heritage, but usually at a more accessible price point. Furthermore, purchasing second-hand luxury items can help to satisfy customers' psychological motivations, such as self-esteem and a sense of belonging, as well as self-actualisation motivations, such as satisfaction and social climbing (Turunen et al., 2020). Indeed, Henninger et al. (2020) advocate that in order to increase sales of sustainable garments, they must appeal to a consumer's self-concept and boost their self-esteem.

Research has identified the two main drivers of second-hand fashion shoppers: environmental and economic factors (Carrigan et al., 2013; Xu et al., 2014; Seo & Kim, 2019; Machado et al., 2019). Economic factors are undoubtedly important as second-hand fashion luxury items are usually lower in price than luxury fashion items, making them more accessible to a wider demographic. On the second-hand luxury platform Vestiare Collective, second-hand luxury fashion items can cost between 30% and 70% less than the original in-store price (Turunen et al., 2020).

However, this is not always the case, as limited-edition second-hand luxury items can cost considerably more (Turunen & Pöyry, 2019). On the other hand, buying second-hand luxury items could be one way that consumers feel like they are engaging in more sustainable consumption as they are purchasing items that already exist as opposed to items that need to be newly produced. Indeed, research suggests that buying second-hand goods can be a source of personal pride for customers who feel that they can make a statement against overconsumption through their purchasing behaviours (Turunen & Leipämaa-Leskinen, 2015). Yet this may also be for personal gain as Henninger et al. (2020) suggest that sustainable fashion purchases have the potential to act as status enhancers, as consumers may want to be seen as being more environmentally friendly and socially responsible. This highlights the complexity of consumers' attitudes towards second-hand luxury fashion and the many reasons that may come into play for purchasing it.

In China, consumers can buy second-hand luxury fashion goods from a wide range of different channels, such as physical stores, e-commerce apps like Hongbulin, and trading platforms such as Xianyu. As of December 2020, luxury second-hand e-commerce customers in China reached 52.66 million, confirming online as the primary sales channel for second-hand luxury goods (CNNIC, 2021). The most popular second-hand luxury products in terms of sales include handbags, watches, sunglasses and jewellery, with women's handbags dominating the sales and accounting for 60.22% of the total market value, or 8.13 billion yuan, in 2019 (CNNIC, 2020). These statistics indicate the growth and rising popularity of the second-hand luxury fashion market in China in recent years. McCormick et al. (2020) found that Chinese Millennials felt a lot of pressure in their lives as they sought to not only please their managers but also fulfil family expectations at being the best, and so purchasing luxury fashion items was a way that they could enjoy life and reward themselves. This indicates that second-hand luxury fashion may be more popular with younger generations in China and that their attitudes towards this industry may differ from older generations.

12.2.2 Social Media in China

China has the largest online retail market (CNNIC, 2020) and is the world's largest social media market (Statista, 2021a). In 2020, there were approximately 926.8 million social media users in China, a number which is forecast to increase to 1.279 billion by 2026 (Statista, 2021c). However, due to the Chinese government's internet censorship (or the 'Great Firewall' of China as it is also known) the majority of Western social media sites, such as Facebook, Twitter and YouTube, are blocked in the country (Statista, 2021a). As a result, a number of alternative social platforms have emerged, such as Little Red Book, Weibo, WeChat and Tiktok.

Weibo functions like Twitter and is a user-relationship-based platform for information exchange (Yao et al., 2019). As one of the top social media sites in China, Weibo accounted for 230 million daily active users in China in the first quarter of 2021, seeing a year-on-year growth of 5 million users (Statista, 2021a). People use Weibo to post a variety of content, including text that is limited to 140 characters, pictures, audio and video content, and other types of media. Little Red Book (or Xiaohongshu, as it is known in China) was founded in 2013 by Miranda Qu and Charlwin Mao and was originally designed as a user-generated social sharing platform. In 2020, Little Red Book had approximately 100 million active users, with 37.2% of them aged between 25 and 34 years (Statista, 2021b). Little Red Book has business at the centre of its design (Yao et al., 2019) and consists of two distinctive functions: the user community (social function) and the e-commerce community (business function).

China has a higher level of collectivism than Western countries (Zhou et al., 2019; Gvili & Levy, 2021), meaning that research findings could differ widely from those based on Western social media platforms. Indeed, Yao et al. (2019) state that social interaction is profoundly integrated into consumers' shopping experiences in China, with social media sites playing a much more prominent role, which distinguishes Chinese consumers from their Western counterparts. Yet the majority of eWoM social media studies have focused on Western social media, with only a few studies investigating Chinese platforms such as Weibo (e.g. Tsai & Men, 2017; Hu et al., 2019) and WeChat (e.g. Chu et al., 2019; Zhang et al., 2021). This study will help to fill this gap.

12.2.3 e-Word-of-Mouth (eWoM)

Word-of-mouth (WoM) is defined as 'oral, person-to-person communication between a perceived non-commercial communicator and a receiver concerning a brand, product, or a service offered for sale' (Arndt, 1967, p. 190). WoM is, therefore, an exchange of information between consumers and can play a pivotal role in shaping consumer behaviour and attitudes towards brands, products and services (Katz & Lazarsfeld, 1966). The development of the internet transformed traditional face-to-face WoM communication to e-Word-of-Mouth (eWoM). Huete-Alcocer (2017) identified four major differences in her comparison of WoM and eWoM: credibility, privacy, diffusion speed and accessibility. eWoM allows consumers to obtain information from a large geographically dispersed group of people who have experienced a product or brand that they are interested in (Zhou et al., 2021). As such, eWoM is very powerful as it enables consumers to obtain, share and communicate about brands and learn more about products that they are interested in in order to help their decision-making process and impact their consumption behaviour (Tien et al., 2019). In particular, Kang et al. (2014) found that fashion-conscious consumers tend to look for opinions about products from eWoM via social media before purchasing. Indeed, social media and eWoM are intrinsically linked, with Chrimes and Boardman (2022) finding that people who search for fashion items on Instagram are also likely to undertake eWoM by sharing fashion products with their friends and commenting on posts themselves. Furthermore, Mohammad et al. (2021) found that eWoM has a direct effect on Malaysian consumers' attitudes towards second-hand clothes, contributing to the increase in consumer acceptance, interest and engagement towards them. This study will explore how eWoM affects Chinese consumers' attitudes and decision-making process in relation to second-hand luxury fashion.

12.2.4 Consumer Decision-Making Process

The purchase decision-making process is a complicated cognitive process including multiple stages of activities that results in choosing and purchasing a product or service from a number of alternatives (Cheung &

Thadani, 2010; Sadovykh et al., 2015). The Engel–Kollat–Blackwell (EKB) model has been widely used as a standard framework to investigate the decision-making process in consumer behaviour research (Darley et al., 2010; Wen et al., 2014). The EKB model shows the five stages that the consumer goes through in order to solve problems and fulfil their needs:

1. The identification of a problem
2. The search for information
3. The development of an alternative
4. A purchase
5. The evaluation of the overall outcome

This model indicates that a purchase decision refers to not only the actual transaction behaviour but also all the cognitive and affective states of the consumer before and after the actual purchase (Engel et al., 1995; Zhang & Benyoucef, 2016). Therefore, all the stages should be considered when studying consumer behaviour instead of focusing on a particular stage (Yadav et al., 2013). This chapter will explore how eWoM on social media may affect the different stages of Chinese consumers' decision-making towards second-hand luxury fashion items through primary data collection. The method of collecting this primary data will now be summarised and discussed.

12.3 Method

An interpretivist qualitative study was conducted in order to explore Chinese consumers attitudes towards second-hand luxury fashion and how eWoM on social media affects their decision-making in relation to second-hand luxury fashion. Semi-structured interviews were conducted with a purposive sample of 24 participants in China, aged 18–35, all of whom had purchased second-hand luxury items in the last 3 months and all of whom were active users of Weibo and Little Red Book. A list of participants, their experience with Weibo/Little Red Book and their interview length can be found in Table 12.1 in the Appendix. Weibo and

Little Red book were chosen as the focus of this study due to their large user bases and the high number of posts relating to fashion items. Five pilot studies were conducted with interviewees who met the sample requirements in order to improve the interview questions by reducing repetitions and increase their intelligibility before data collection began. The study was exploratory and aimed to elicit an in-depth understanding of consumers' attitudes and behaviour towards second-hand luxury fashion and social media eWoM (Boardman & McCormick, 2018). Open-ended questions were asked whereby participants were encouraged to answer by explaining their thoughts in their own words. For this reason, the interview questions started with 'how' or 'what' in order to encourage participants to provide more information (Creswell, 2009). Participants were asked 'How would you define second-hand luxury fashion items?' A general discussion followed, whereby luxury fashion items were defined as any clothes, handbags or accessories that were created by recognised premium brands such as Louis Vuitton, Prada, Gucci and Bottega Veneta. Second-hand was defined as any item that has been pre-owned. Participants were also asked 'What is your understanding of e-word-of-mouth on social media?' and 'What would you identify as eWoM?' A general discussion clarifying what the interviewer meant by eWoM and examples followed in order to ensure that each participant could correctly identify eWoM. In order to address RQ1 and RQ2 throughout the interview, participants were asked questions such as 'What's your overall attitude towards second-hand luxury fashion?', 'What could make you change your attitude towards second-hand luxury fashion?' or 'How do you feel when you see eWoM relating to second-hand luxury fashion?', 'What do you think about this kind of post?', 'How might that affect your decision-making when it comes to a second-hand luxury item?' Saturation point was reached. The interviews were transcribed in Chinese verbatim and then translated into English. The researchers conducted multiple coding cycles, first individually, and then collectively, discussing any discrepancies of emerging themes and patterns. The findings from the data analysis are presented below.

12.4 Findings

This study explored Chinese consumers' attitudes towards second-hand luxury fashion and how eWoM on social media may affect their decision-making towards them. The following section will analyse and discuss the findings from the semi-structured interviews, framed by the research questions.

RQ1: What are the Attitudes of Chinese Consumers to second-hand luxury fashion?

Data showed that **price, symbolic value/status, uniqueness, re-sale value and sustainability** were the most common reasons for purchasing second-hand luxury fashion amongst Chinese consumers. However, **hygiene, lack of traceability, signs of wear** and doubts surrounding **authenticity** were reasons why Chinese consumers may be put off by second-hand luxury fashion. Chinese consumers' attitudes towards second-hand luxury fashion will now be discussed in depth.

12.4.1 Price

Price was mentioned by several participants as the main reason for purchasing second-hand luxury fashion. Participants perceived luxury fashion to be very expensive and so second-luxury items enabled them to own luxury items when otherwise they would not have been able to afford to, as summarised by participant 23:

> For those who want to buy a luxury handbag but don't have a big enough budget, second-hand luxury is a good option. (Participant 23)

Thus, second-hand luxury fashion made **luxury more accessible** to Chinese consumers due to its **affordability**. However, some participants saw the lower prices and increased accessibility as a negative thing, as it **reduced luxury brands' exclusivity**. On the other hand, some participants expressed anger at the fact that sometimes second-hand luxury fashion could cost as much as new luxury fashion items, for example:

Some sellers will use the scarcity of products to increase prices and this makes me feel negatively towards second-hand luxury. (Participant 16)

This indicates that second-hand luxury fashion items must remain more affordable than new luxury items in order for it to receive favourable attitudes amongst Chinese consumers but that there is a conflict of interest here as the more affordable items are, the less exclusive they are, which diminishes their brand value.

12.4.2 Symbolic Value/Status

Participants mentioned the **symbolic value** of luxury as one of the advantages of buying second-hand luxury. This value is conveyed by the display of social status associated with luxury brands, exemplified by Participant 10:

I might purchase second-hand luxury because of how other people will then view me … It makes me feel confident and that I can earn the respect or get attention from others for less cost compared with purchasing new luxury. (Participant 10)

This implies that despite the items being pre-owned, they are still perceived as luxury, and are viewed as an effective alternative to showcase status and symbolic belonging to idealised social groups.

12.4.3 Uniqueness and Self-Expression

Another reason that Chinese consumers liked to buy second-hand luxury fashion items was because of their **uniqueness**. For example:

I can buy some products that are no longer available, such as limited editions. (Participant 20)

This suggests that Chinese consumers liked to purchase second-hand luxury due to its rarity and ability to help them stand out from the crowd as a way of **self-expression**. Indeed, participant 1 said:

Nowadays, the popularity of luxury styles in China has become pervasive ... on social media, you basically see people carrying the same bag on the street. I think this is a bit embarrassing. There are too many people with the same bag. I want to find some niche designs, it doesn't matter if it's older. (Participant 1)

This implies that some Chinese consumers, namely fashion leaders, view modern luxury fashion items as being more mass-produced and do not want to be seen wearing the same styles as everyone else. Fashion leaders use second-hand luxury handbags to show their fashion style and identity in a more unique way, not wishing to look the same as everyone else. This is interesting as previous literature discusses the strong influence of collectivism in Chinese consumers (e.g. Zhou et al., 2019; Gvili & Levy, 2021). Yet, the findings suggest that some Chinese consumers who purchase second-hand luxury do so in order to display their own unique style. Participants liked to own something that is unique that their peers do not own, as expressed by Participant 2:

Second-hand luxury is more limited and it's harder for people to find the same items, except for the classic styles maybe ... It is cool that I can wear luxury but no one has seen this style. (Participant 2)

Thus, second-hand luxury provides a mechanism for self-expression whereby consumers are able to showcase their unique, personal style by wearing items that are not readily available to be purchased by lots of others, creating a sense of exclusivity in another way.

12.4.4 Re-sale Value

Some participants said that the re-sale value was a reason that they purchased second-hand luxury goods. This was particularly true for second-hand luxury handbags, as demonstrated by Participant 17:

If you want to sell them again later, there is still a good market for some of the more valuable brands of second-hand luxury handbags. (Participant 17)

Some participants sought out exclusive limited-edition second-hand luxury items purely to re-sell them at a higher price to make it a profitable enterprise, for example:

> For some limited editions, the value is even greater than when they were first released in the secondary market. (Participant 18)

Thus, for some of the rarer items, buying second-hand was seen as an investment opportunity, as some limited-edition styles were re-sold at higher prices.

12.4.5 Sustainability

Participants also discussed sustainability as a reason for purchasing second-hand luxury fashion:

> Second-hand luxury handbags are more environmentally friendly. (Participant 15)

Participants believed that by purchasing second-hand luxury they were acting in a more responsible way that was better for the environment as they 'reduce waste' (Participant 1). However, it was acknowledged that this was not the primary reason for purchasing second-hand luxury, with price and symbolism/status being seen as much more important to Chinese consumers.

12.4.6 Authenticity Doubts

Despite the overall positive attitudes expressed towards second-hand luxury fashion items by Chinese consumers, there were some concerns shared. Interview data revealed that one of the biggest reasons why Chinese consumers are wary of purchasing second-hand luxury fashion is that they have **doubts about the authenticity** of the items. This is summarised by Participant 19:

Fakes are prevalent in the market; I am very worried about buying fake second-hand bags. (Participant 19)

Seeing stories on social media about fake second-hand luxury handbags instilled a fear in other potential customers, discouraging them from buying second-hand luxury fashion. This is important as Chinese consumers wanted the items to be genuine as they wanted the pre-owned items to convey the same meaning and authenticity as brand-new luxury goods. One of the main reasons for these fears was due to the lack of regulations in the Chinese second-hand luxury market, as discussed by Participant 24:

I think buying and using second-hand luxury handbags is very environmentally friendly and can help me save money. On the other hand, I am very afraid of buying fakes, because the certification and sales of the Chinese second-hand market are not yet very mature and regulated. (Participant 24)

This fear of second-hand luxury items being counterfeits appears to be more prevalent amongst Chinese consumers due to a perception of a lack of regulation and standardisation of the Chinese second-hand luxury fashion market. There is a lack of discussion regarding doubt over authenticity in studies conducted in other countries in the literature, suggesting that there are regional and cultural differences in attitudes towards second-hand luxury fashion.

12.4.7 Hygiene

Some participants were put off second-hand luxury fashion due to potential hygiene issues, for example:

I am conscious about the cleanliness of the used items. (Participant 1)

However, handbags were seen to be more immune to this than clothes as they are not worn on the body, as discussed by Participant 17:

Second-hand luxury handbags are a good option, after all, bags are not like clothes that you wear around, they are still fine from a hygiene point of view. And most of the second-hand luxury handbags I've seen so far are still in good condition. (Participant 17)

Yet, interestingly, some participants alluded to the fact that positive eWoM had changed their mind about the hygiene of second-hand items:

I actually couldn't accept products that had been used by others at the beginning, but after reading other people's reviews and bloggers' recommendations on social networks, especially some positive experiences, I now think that second-hand luxury items are a good choice. (Participant 20)

This highlights the power of eWoM on social media and how it can influence Chinese consumers' views on second-hand luxury fashion.

12.4.8 Lack of Traceability

Some participants did not like the idea that someone that they did not know had previously owned the item:

I don't know what happened to the last user. I think that items have memories. Strangers have used them. They are old. I'm afraid they were carried by criminals or other people that I don't know. (Participant 1)

On the other hand, some participants liked the fact that the items had a history:

Second-hand stores sell many items that have stories, but they don't write the story. Suppose a ring is the best thing that a man can buy when he marries his wife. He spent many years with his wife and the ring stayed with her until now. This is what I want to buy ... it's the memories and the meanings that comes with it ... I think that second-hand luxury is not a simple commodity, but a cultural heritage. I will pay great attention to the story behind the item. (Participant 7)

This indicates that further transparency about the history of second-hand luxury fashion items would be appealing to Chinese consumers, both by putting some people at ease so that they knew who had owned the item but also for consumers that are interested in owning items that have had an interesting previous life and want to know their stories.

12.4.9 Signs of Wear

Participants discussed the main drawback of second-hand luxury bags as the fact that they have **signs of wear**. This is summarised by Participant 21:

> Second-hand luxury goods will have use traces, such as scratches and fading on the hardware of luxury bags. (Participant 21)

Participants discussed how signs of wear make second-hand luxury items easily **distinguishable from brand-new items** which **lessened their value**. Indeed, participants were dismayed at the fact that sometimes the signs of wear 'will be very visible' especially if it 'has not been well cared for' (Participant 18). This shows that the condition of the second-hand luxury fashion item is very important to Chinese consumers and will affect whether their attitude is favourable to items or not, as signs of wear may diminish the status that consumers want to portray.

RQ2: How does eWoM on social media affect Chinese consumers' decision-making in relation to second-hand luxury fashion?

The findings for RQ2 explored how eWoM affects the five stages of Chinese consumers' decision-making process and are discussed below.

1. **eWoM and the Identification of a Problem**

The primary trigger encouraging Chinese to want to buy second-hand luxury fashion items was seeing **influencers/Key Opinion Leaders (KOLs)** on social media wearing second-hand luxury items. The following quote summarises how eWoM by influencers affects Chinese consumers:

...As I gradually scroll through Little Red Book, I see celebrities and bloggers recommending attractive second-hand luxury items. Usually the prices are not too high, but the styles are very unique. This makes me realise that I also need to buy second-hand luxury! (Participant 3)

This indicates that the most influential source of eWoM on social media for Chinese consumers is that which is delivered by influencers and Key Opinion Leaders (KOLs).

2. eWoM and Information Search

Participants mentioned that they scrolled through Little Red Book as a way to relax, as well as using it as a tool for searching for items and to gain inspiration. The interviews revealed that the most frequently used function was the 'Auto Recommendation' function, whereby Little Red Book recommends the most suitable content for each user based on their previous search behaviours. Participants mentioned that Little Red Book effectively organised product classifications and they tended to search category-specific information on there. As such, Chinese consumers found Little Red Book to be a **convenient** platform to search for information about second-hand luxury fashion products:

Little Red Book helps me to sum up what different people are sharing about their purchase and sometimes I will go and explore that item further, so I think it's very convenient. (Participant 12)

Participants evaluated the quality of eWoM based on the presence of **original and creative content**. Interviews revealed that high-quality information from **consumers who had experienced products** improved Chinese consumers' perceptions of second-hand luxury items. It was important to consumers that the posts were original, as 'if they copy someone else's content, then this information is meaningless to me' (Participant 17). Thus, Chinese consumers wanted to see social media posts relating to individuals' 'own personal experience' with second-hand products in order to help them with their 'understanding of the product' (Participant 22).

12 Chinese Consumer Attitudes Towards Second-Hand... 257

Participants said that the inclusion of **pictures** aided their decision-making because pictures reflected the characteristics of the product more intuitively than words, as discussed by Participant 19:

> If a post about a product is accompanied by a real and good-looking picture, it will attract me. (Participant 19)

Thus, the interviews revealed that eWoM in the form of attractive and 'real' imagery was more eye-catching and more helpful in terms of decision-making for Chinese consumers.

Participants indicated that the nature of the sender determined the credibility of the eWoM. If it is posted by a **blogger** that the consumer follows and whose opinion they trust, then the information was considered **credible**, as illustrated by Participant 19:

> I prefer information spread by popular bloggers, and I will pay more attention to what they say in relation to products. (Participant 19)

Yet, participants expressed that the amount of information available on social media meant that it required sorting and processing, which was time and energy consuming. This resulted in **information overload**. The negative effect that this had on consumers was summarised by Participant 12:

> I used to read the comments on Little Red Book before I bought anything, both good and bad comments, and I compared them myself. But now I don't. I think that is such as waste of energy and I just prefer going shopping directly ... because I don't want all my spare time being spent on my mobile phone. Shopping should be a happy and relaxing thing. (Participant 12)

The study found that although consumers found it convenient to collect information on second-hand luxury fashion items on Little Red Book, due to the excessive amount of information on there, consumers found it difficult to digest it all. The interview data suggests that the more information that there was about items and the longer the time consumers spent looking at eWoM on social media, the likelihood of consumers making an impulse purchase diminished.

3. eWoM and the Development of Alternatives

Consumers compared the information that they gathered about second-hand luxury fashion items in order to help them decide whether to make a purchase or not. Consumers found the **comments** to be influential during this stage of their decision-making process, as illustrated by Participant 10:

> I will often see many people ask others in the comment area to help them select which bag is better and I will compare them and then finally make up my mind in terms of which bag I want to buy, and then I will find various channels to buy it. (Participant 10)

Participants mentioned that they are more likely to be influenced by eWoM containing an **evaluation of a product**. This is because consumers felt that evaluations of products 'can help me avoid buying things that are not suitable for me' (Participant 21). More specifically, if the eWoM evaluated the product material, it seemingly enhanced the credibility of the information as it was deemed to be reliable. Indeed, many participants were **wary of the credibility of the eWoM** on social media, making them quite distrustful of many posts. The presence of other consumers' positive comments helped to support the credibility of the original post, whether they related to the eWoM message or the product mentioned. Therefore, **consumers read the comments in order to reassure themselves of the eWoM's authenticity and reliability**. For example:

> I would look for opinions about this information in the comments. If there are many opinions that agree with it, the credibility will be higher. (Participant 17)

However, a high number of comments relating to specific second-hand luxury fashion items did not enhance consumers' trust of the eWoM regarding it. In fact, if there was a sudden rush of reviews for the same product, consumers were more sceptical, believing it to be marketing strategy implemented by the seller, not an authentic review posted by a real consumer.

4. eWoM and the Purchase

The interviews indicated that **positive eWoM was likely to increase consumers' willingness to buy second-hand luxury fashion items**, as illustrated by Participant 13:

> …Because I have limited access to information related to second-hand luxury bags, when there is positive eWoM related to it, I will refer to it and if it is positive then I would buy it. (Participant 13)

eWoM was particularly influential due to the lack of other information sources relating to second-hand luxury fashion items. However, the **eWoM was more influential if consumers already had an item in mind** beforehand.

Conversely, the exposure to **negative eWoM made some participants more reluctant to buy second-hand luxury fashion items**, summarised by Participant 16:

> …if I see negative eWoM, I will definitely not buy second-hand luxury items. I would worry about buying second-hand items then. (Participant 16)

This was particularly true if the **negative eWoM related to items being fake**, as participants were then completely put off buying second-hand luxury items then, illustrated by participant 15:

> If I see lots of negative reviews saying, for example, 'the bag I bought is a fake', it will make me worry about whether I will buy fakes if I buy second-hand luxury bags and completely change my attitude about it. (Participant 15)

Furthermore, as most consumers purchased second-hand luxury online, one of the main worries was **whether the images were a true** reflection of the product. This is true for any online purchases but was particularly pertinent for second-hand items, due to the variations in condition, as discussed by Participant 17:

> I will be wondering whether the information of this bag is true ... thinking is there a big difference between the second-hand luxury and the first-hand one in terms of condition? (Participant 17)

In order to help them overcome the risk, consumers turned to eWoM about the seller and the item:

> I will consider whether the pictures match the actual handbag, and whether the person who posted this information is credible. (Participant 20)

The credibility of the eWoM was of great concern to the participants, they needed to be able to trust the source in order to be able to trust the eWoM. This is because consumers found it difficult to determine the authenticity of a product and were worried about items being fake.

Once the decision to purchase the item was made, consumers selected the appropriate channel to make the purchase. WeChat was the most common way for participants to buy second-hand luxury, but this in itself raised concerns, for example:

> The sales channel of second-hand luxury from Little Red Book often entails asking the customer to add the seller's private WeChat, which is very unofficial. I think it is risky. (Participant 1)

This highlights the worries that Chinese consumers have over the purchasing of second-hand luxury goods as a result of the lack of professional channels selling items.

5. eWoM and the Evaluation of the Overall Outcome

The evaluation of the overall outcome includes any feelings or actions that occur after the purchase has been made and the item has been received. A concern expressed by all participants when purchasing second-hand luxury fashion items in China related to the authenticity of products. Some participants indicated that the process was daunting:

> Different people say different things about the reputations of various channels on Little Red Book when it comes to purchasing second-hand luxury. Some say they bought real ones, and some say they bought fake ones. It makes me so conscious of paying for products. I don't know whether I will be lucky or not. (Participant 1)

This highlights the levels of distrust felt by consumers in the Chinese second-hand luxury fashion market in relation to the authenticity of items and the belief that many items being sold were fake. Some customers even took items to department stores for authenticity checks once they'd received it.

In summary, the main concerns expressed by Chinese consumers when purchasing second-hand luxury fashion were **channel transparency, price, the authenticity of the product and communication with the seller** or the trading platforms.

12.5 Conclusion

This study found that price, symbolic value/status, uniqueness, re-sale value and sustainability were the most common reasons for purchasing second-hand luxury fashion amongst Chinese consumers. Second-hand luxury fashion items must remain more affordable than new luxury items in order to receive favourable attitudes amongst Chinese consumers as lower prices were perceived to be its main advantage. Chinese consumers also bought second-hand luxury items for their uniqueness in order to stand out from the crowd in a saturated luxury fashion market. In particular, this study provides novel findings highlighting the connection between second-hand luxury fashion and its use by fashion leaders to showcase their own unique style and identity, as second-hand items are rarer and not easily purchased. In post-modern society, consumption is the main means of constructing identity (Koksal, 2013), and this research suggests that Chinese consumers feel like they can do this through purchasing second-hand luxury fashion items as the items maintain the same level of symbolism as new luxury items but the specific items are scarcer than new mass luxury items. Yet, this could change if the lower prices of second-hand luxury items increased accessibility to luxury brands, in general, as it reduced the level of exclusivity and desirability of the brand itself. Furthermore, the condition of the second-hand luxury fashion item is very important to Chinese consumers and affected whether their attitude was favourable towards second-hand luxury items or not, as any signs of wear may diminish the status that consumers want to portray when wearing the items. Consumers also considered the re-sale value to be a major benefit of purchasing second-hand luxury items, particularly handbags. Although some participants mentioned sustainability as a reason for purchasing second-hand luxury fashion, this was not the primary reason for doing so, with price and symbolism being seen as much more important to Chinese consumers.

Data revealed that doubts over the authenticity of items, hygiene, a lack of traceability and signs of wear were the reasons why Chinese consumers had unfavourable attitudes towards second-hand luxury fashion. In particular, some participants did not like the idea that someone that they did not know had previously owned the item. Future research could investigate whether more transparency and details over previous ownership could help solve this barrier to second-hand luxury fashion and whether new technologies such as blockchain could help facilitate this. Indeed, some consumers liked the idea that each item had its own story and were part of cultural history, and so further transparency may help to highlight these stories too. Moreover, technology could help remove one of the other barriers in not knowing whether items are genuine or fakes.

This research explores how eWoM on social media affects the five stages of Chinese consumers' decision-making process. The study found that the key trigger for prompting second-hand luxury purchases and the most influential source of eWoM on social media for Chinese consumers was that delivered by influencers/Key Opinion Leaders (KOLs). This was due to their fashion leader status, but also because they were seen as more credible and trustworthy sources of second-hand luxury fashion information. Chinese consumers found it convenient to collect information on second-hand luxury fashion items on social media, but the excessive amount of information over there resulted in information overload. Chinese consumers were more likely to be influenced by eWoM containing an evaluation of a product, particularly if it related to the material, and posts that contained 'real-life' imagery as they were more helpful in aiding their decision-making process. Furthermore, as most consumers purchased second-hand luxury online, one of the main worries was whether the images were a true reflection of the product, which was particularly important due to the variations in the condition of second-hand items. eWoM about the seller and the item posted on social media helped to reassure consumers and overcome this risk to a certain extent.

However, many participants were wary of the credibility of eWoM on social media, making them quite distrustful of many posts. This was because consumers found it difficult to distinguish the authenticity of products and were worried about items being fake. The presence of other consumers' positive comments helped to support the credibility of the original post, whether they related to the eWoM message or the product mentioned. Therefore, consumers read the comments in order to reassure themselves of the eWoM's authenticity and reliability. In contrast to previous findings (e.g. Matute et al., 2016), they cared less about the quantity of eWoM, as large numbers of positive reviews made them doubt their authenticity. Therefore, the study found that the more information there was on items and the longer the time spent on social media, the likelihood of an impulse purchase diminished. Indeed, if there was a sudden rush of reviews for the same product, consumers were more sceptical, believing it to be marketing strategy implemented by the seller, not an authentic review posted by a real consumer.

In line with existing literature, the study found that positive eWoM was likely to increase Chinese consumers' willingness to buy second-hand luxury fashion items and that negative eWoM had the opposite effect. eWoM was particularly influential on consumers' decision-making due to the lack of other information sources relating to second-hand luxury fashion available. However, eWoM was more influential if consumers already had an item that they wanted to buy in mind beforehand. If the negative eWoM related to items being fake, then consumers were put off second-hand luxury fashion items completely, highlighting the importance of authenticity in relation to second-hand luxury fashion in China. Some customers even took items to department stores for authenticity checks after they had purchased it. This highlights the levels of distrust felt by Chinese consumers regarding the Chinese second-hand luxury fashion market in relation to the authenticity of items and the belief that many items being sold were fake. One of the main reasons for these fears was due to the lack of regulations in the Chinese second-hand luxury market, an aspect that was discussed by many participants. Therefore, this research suggests there may be regional and cultural differences with attitudes towards second-hand luxury with more doubt and fears about counterfeiting and the authenticity of the products amongst Chinese consumers than other countries and cultures.

These findings have implications for luxury brands as well as second-hand luxury sellers. If luxury brands choose to set up a second-hand sales platform of their own, something which is likely to happen in the future due to the rising success of re-sale platforms creating a new form of competition for luxury fashion brands, the second-hand luxury fashion items need to be more affordable than new luxury items in order to receive favourable attitudes amongst Chinese consumers. This study found that Chinese consumers buy luxury and second-hand luxury fashion in order to help construct their identity. Luxury items must, therefore, be unique in terms of design as Chinese consumers want to stand out from the crowd in a saturated luxury fashion market. Finally, Chinese consumers were more likely to be influenced by eWoM containing an evaluation of a product, particularly if it related to the material, and posts that contained 'real-life' imagery as they were more helpful in aiding their decision-making process. Luxury brands can use this to inform their social media marketing strategies.

Appendix

Table 12.1 Interview information

Interviewee	Little red book/weibo user?	Interview length
Participant 1	Weibo & Little Red Book User	33 min
Participant 2	Weibo & Little Red Book User	45 min
Participant 3	Weibo & Little Red Book User	32 min
Participant 4	Weibo & Little Red Book User	35 min
Participant 5	Weibo & Little Red Book User	30 min
Participant 6	Weibo & Little Red Book User	52 min
Participant 7	Weibo & Little Red Book User	40 min
Participant 8	Weibo & Little Red Book User	35 min
Participant 9	Weibo & Little Red Book User	30 min
Participant 10	Weibo & Little Red Book User	42 min
Participant 11	Weibo & Little Red Book User	55 min
Participant 12	Weibo & Little Red Book User	38 min
Participant 13	Weibo & Little Red Book User	54 min
Participant 14	Weibo & Little Red Book User	37 min
Participant 15	Weibo & Little Red Book User	32 min
Participant 15	Weibo & Little Red Book User	51 min
Participant 17	Weibo & Little Red Book User	48 min
Participant 18	Weibo & Little Red Book User	30 min
Participant 19	Weibo & Little Red Book User	45 min
Participant 20	Weibo & Little Red Book User	35 min
Participant 21	Weibo & Little Red Book User	30 min
Participant 22	Weibo & Little Red Book User	35 min
Participant 23	Weibo & Little Red Book User	38 min
Participant 24	Weibo & Little Red Book User	30 min

References

Arndt, J. (1967). Word-of-mouth advertising and informal communication. In M. A. Boston (Ed.), *Risk taking and information handling in consumer behavior* (pp. 188–239). Harvard University Press.

Bain & Company. (2021). The Future of Luxury: Bouncing Back from Covid-19. Available from: https://www.bain.com/insights/the-future-of-luxurybouncing-back-from-covid-19/

Banbury, C., Stinerock, R., & Subrahmanyan, S. (2012). Sustainable consumption: Introspecting across multiple lived cultures. *Journal of Business Research, 65*(4), 497–503.

Boardman, R., & McCormick, H. (2018). Shopping channel preference and usage motivations: Exploring differences amongst a 50-year age span. *Journal of Fashion Marketing and Management: An International Journal, 22*(2), 270–284.

Carrigan, M., Moraes, C., & McEachern, M. (2013). From conspicuous to considered fashion: A harm-chain approach to the responsibilities of luxury-fashion businesses. *Journal of Marketing Management, 29*(11–12), 1277–1307.

Cheung, C. M. K., & Thadani, D. R. (2010). The effectiveness of electronic word-of-mouth communication: A literature analysis. *Bled eConference, 23,* 329–345.

Chevalier, M., & Mazzalovo, G. (2012). *Luxury Brand Management a world of privilege* (2nd ed.). Wiley.

Chrimes, C., & Boardman R. (2022). Investigating shopper motivations for purchasing on Instagram. *International Journal of Technology Transfer and Commercialisation,* special issue on social commerce (accepted in press).

Chu, S., Lien, C., & Cao, Y. (2019). Electronic word-of-mouth (eWOM) on WeChat: Examining the influence of sense of belonging, need for self-enhancement, and consumer engagement on Chinese travellers' eWOM. *International Journal of Advertising, 38*(1), 26–49.

CNNIC. (2020). *Statistical report on internet development in China, China Internet Network Information Center.* Retrieved from https://www.cnnic.com.cn/IDR/ReportDownloads/202012/P020201201530023411644.pdf

CNNIC. (2021). *The 47th China statistical report on internet development.* Retrieved from http://www.cnnic.net.cn/hlwfzyj/hlwxzbg/hlwtjbg/202102/P020210203334633480104.pdf

Creswell, J. (2009). *Research design: Qualitative, quantitative and mixed methods approaches* (3rd ed.). Sage.

Darley, W. K., Blankson, C., & Luethge, D. J. (2010). Toward an integrated framework for online consumer behavior and decision-making process: A review. *Psychology and Marketing, 27*(2), 94–116.

Diderich, J., & Theodosi, N. (2021, September 1). *Why luxury brands are sitting out the resale market boom, WWD.* Retrieved from https://wwd.com/fashion-news/designer-luxury/luxury-brands-reluctant-to-join-resale-market-1234898376/

Engel, J. F., Blackwell, R. D., & Miniard, P. W. (1995). *Consumer behavior* (8th ed.). Harcourt Education.

Fionda, A. M., & Moore, C. M. (2009). The anatomy of the luxury fashion brand. *The Journal of Brand Management, 16*(5–6), 347–363.

Gao, L., Norton, M. J. T., Zhang, Z., & Kinman To, C. (2009). Potential niche markets for luxury fashion goods in China. *Journal of Fashion Marketing and Management, 13*(4), 514–526.

Gopalakrishnan, S., & Matthews, D. (2018). Consumer attitudes and communication in circular fashion. *Journal of Fashion Marketing and Management, 22*(3), 189–208.

Gvili, Y., & Levy, S. (2021). Consumer engagement in sharing brand-related information on social commerce: The roles of culture and experience. *Journal of Marketing Communications, 27*(1), 53–68.

Henninger, C. E., Blazquez, M., Boardman, R., Jones, C., McCormick, H., & Sahab, S. (2020). Cradle-to-cradle versus consumer preferences in the fashion industry. In S. Hashmi & A. I. Choudhury (Eds.), *Encyclopaedia of renewable and sustainable materials, 5* (pp. 353–357). Elsevier.

Hu, X., Chen, X., & Davison, R. M. (2019). Social support, source credibility, social influence, and impulsive purchase behavior in social commerce. *International Journal of Electronic Commerce, 23*(3), 297–327.

Huete-Alcocer, N. (2017). A literature review of word of mouth and electronic word of mouth: Implications for consumer behavior. *Frontiers in Physiology, 8*, 1256–1256.

Kang, J. Y. M., Johnson, K. K. P., & Wu, J. (2014). Consumer style inventory and intent to social shop online for apparel using social networking sites. *Journal of Fashion Marketing and Management, 18*(3), 301–320.

Katz, E., & Lazarsfeld, P. F. (1966). *Personal influence: The part played by people in the flow of mass communications.* Transaction Publishers.

Koksal, M. H. (2013). Psychological and behavioural drivers of male fashion leadership. *Asia Pacific Journal of Marketing and Logistics, 26*(3), 430–449.

Machado, M. A. D., Almeida, S. O. D., Bollick, L. C., & Bragagnolo, G. (2019). Second-hand fashion market: Consumer role in circular economy. *Journal of Fashion Marketing and Management, 23*(3), 382–395.

Matute, J., Polo-Redondo, Y., & Utrillas, A. (2016). The influence of EWOM characteristics on online repurchase intention: Mediating roles of trust and perceived usefulness. *Online Information Review, 40*(7), 1090–1110.

McCormick, H., Zhang, R., Boardman, R., Jones, C., & Henninger, C. E. (2020). 3D printing in the fashion industry: A fad or the future? In G. Vignali, L. Reid, D. Ryding, & C. E. Henninger (Eds.), *Technology-drive sustainability: Innovation in the fashion supply chain* (pp. 137–154). Palgrave Macmillan.

Mohammad, J., Quoquab, F., & Mohamed Sadom, N. Z. (2021). Mindful consumption of second-hand clothing: The role of Ewom, attitude and consumer engagement. *Journal of Fashion Marketing and Management, 25*(3), 482–510.

Roux, D., & Guiot, D. (2008). Measuring second-hand shopping motives, antecedents and consequences. *Recherche et Applications en Marketing, 23*(4), 64–84.

Sadovykh, V., Sundaram, D., & Piramuthu, S. (2015). Do online social networks support decision-making? *Decision Support Systems, 70*, 15–30.

Seo, M. J., & Kim, M. (2019). Understanding the purchasing behaviour of second-hand fashion shoppers in a non-profit thrift store context. *International Journal of Fashion Design, Technology and Education, 12*(3), 301–312.

Statista. (2021a). Retrieved from https://www.statista.com/topics/1170/social-networks-in-china/#dossierKeyfigures

Statista. (2021b). *Active user age distribution of Xiaohongshu in 2020*. Retrieved from https://www.statista.com/statistics/1053545/china-xiaohongshu-user-age-distribution/

Statista. (2021c). Number of social network users in China from 2017 to 2020 with a forecast until 2026, Available at: https://www.statista.com/statistics/277586/number-of-social-network-users-in-china/ [accessed 30.06.22].

Tien, D. H., Amaya Rivas, A. A., & Liao, Y.-K. (2019). Examining the influence of customer-to-customer electronic word-of-mouth on purchase intention in social networking sites. *Asia Pacific Management Review, 24*(3), 238–249.

Tsai, W.-H. S., & Men, L. R. (2017). Consumer engagement with brands on social network sites: A cross-cultural comparison of China and the USA. *Journal of Marketing Communications, 23*(1), 2–21.

Turunen, L. L. M., & Leipämaa-Leskinen, H. (2015). Pre-loved luxury: Identifying the meanings of second-hand luxury possessions. *The Journal of Product & Brand Management, 24*(1), 57–65.

Turunen, L. L. M., & Pöyry, E. (2019). Shopping with the resale value in mind: A study on second-hand luxury consumers. *International Journal of Consumer Studies, 43*(6), 549–556.

Turunen, L. L. M., Cervellon, M.-C., & Carey, L. D. (2020). Selling second-hand luxury: Empowerment and enactment of social roles. *Journal of Business Research, 116*, 474–481.

Wen, C., Prybutok, R., Blankson, C., & Fang, J. (2014). The role of Equality within the consumer decision-making process. *International Journal of Operations & Production Management, 34*(12), 1506–1536.

Xu, Y., Chen, Y., Burman, R., & Zhao, H. (2014). Second-hand clothing consumption: A cross-cultural comparison between American and Chinese young consumers. *International Journal of Consumer Studies, 38*(6), 670–677.

Yadav, M. S., de Valck, K., Hennig-Thurau, T., Hoffman, D. L., & Spann, M. (2013). Social commerce: A contingency framework for assessing marketing potential. *Journal of Interactive Marketing, 27*(4), 311–323. https://doi.org/10.1016/j.intmar.2013.09.001

Yao, Y., Boardman, R., & Vazquez, D. (2019). Cultural considerations in social commerce: The differences and potential opportunities in China. In R. Boardman, M. Blazquez, C. Henninger, & D. Ryding (Eds.), *Social commerce: Consumer behaviour in online environments* (pp. 43–58). London.

Zhang, K. Z. K., & Benyoucef, M. (2016). Consumer behavior in social commerce: A literature review. *Decision Support Systems, 86*, 95–108.

Zhang, L., & Cude, B. J. (2018). Chinese consumers. Purchase intentions for luxury clothing: A comparison between luxury consumers and non-luxury consumers. *Journal of International Consumer Marketing, 30*(5), 336–349.

Zhang, H., Liang, X., & Qi, C. (2021). Investigating the impact of interpersonal closeness and social status on electronic word-of-mouth effectiveness. *Journal of Business Research, 130*, 453–461.

Zhou, S., McCormick, H., Blazquez, M., & Barnes, L. (2019). eWOM: The rise of the opinion leaders. In R. Boardman, M. Blazquez, C. Henninger, & D. Ryding (Eds.), *Social commerce*. Palgrave Macmillan.

Zhou, S., Barnes, L., McCormick, H., & Blazquez Cano, M. (2021). Social media influencers' narrative strategies to create Ewom: A theoretical contribution. *International Journal of Information Management, 59*, 102293.

13

The Rise of Virtual Representation of Fashion in Marketing Practices: How It Can Encourage Sustainable Luxury Fashion Consumption

Shuang Zhou, Eunsoo Baek, and Juyeun Jang

13.1 Introduction

The strategic use of digital technologies in marketing efforts is well documented for luxury fashion brands (Javornik et al., 2021; Holmqvist et al., 2020). Much of prior work focuses on social media that provides a virtual space for consumers and brands to co-create value and improve brand equity (Bazi et al., 2020). Beyond social media, luxury fashion brands are increasingly deploying immersive technologies to digitise their goods and services (Morewedge et al., 2021). Notable examples are producing

---

S. Zhou (✉) • J. Jang
School of Fashion and Textiles, The Hong Kong Polytechnic University, Hung Hom, Kowloon, Hong Kong, China
e-mail: amysh.zhou@polyu.edu.hk; juyeun.jang@polyu.edu.hk

E. Baek
Department of Clothing & Textiles, Hanyang University, Seoul, South Korea
e-mail: ebaek@hanyang.ac.kr

© The Author(s), under exclusive license to Springer Nature Switzerland AG 2022
C. E. Henninger, N. K. Athwal (eds.), *Sustainable Luxury*, Palgrave Advances in Luxury, https://doi.org/10.1007/978-3-031-06928-4_13

virtual goods, using virtual stores, and adopting virtual reality (VR) and augmented reality (AR) in marketing communications to promote sustainable products (Javornik et al., 2021; Deloitte, 2020). These technologies provide unique capacity characteristics for luxury fashion brands to blur the line between the virtual and physical worlds and reduce materials waste (Deloitte, 2020). While existing research considers the adoption of digital technologies to build brand equity, explorations on how digital technologies influence sustainable luxury consumption are sparse. This chapter proposes that embedding digital technologies in luxury fashion marketing contributes to the virtual representation of fashion (VRF). VRF is defined as digitising material and non-material things of fashion brands to create and display visual counterparts in a virtual space or a space where the physical and the virtual coexist. Here, digital technologies are about: (a) electronic tools such as AR, VR, artificial intelligence (AI) and 2D/3D modelling that create computer-generated information and simulation related to brands; and (b) tools, media, systems, or platforms for delivering the information and simulation to users, such as social media, mobile apps, online games, live streaming, Internet of things (Ameen et al., 2021). With sustainable potentials, VRF could provide vast scope in investigating the approaches to enhance sustainable luxury fashion consumption (SLFC).

Luxury fashion goods are perceived as superior craftsmanship and aesthetics, prestige pricing, and uniqueness. They are frequently associated with endurance, timelessness, exclusivity, social status display, and recognition, differentiating them from fast fashion goods that encourage consumers to purchase short-lived, disposable clothing and accessories (Zhang & Zhao, 2019). In this regard, luxury consumption seems to be somewhat compatible with sustainable consumption as both are grounded on the core principles of rarity and durability that feature consuming less (Jung et al., 2021). However, in the last two decades, the 'massification' of luxury boosted the prevailing of entry-level products and services (Silverstein & Fiske, 2003). Given the ease of accessibility of luxury brands at affordable prices, demand for luxury fashion goods has increased globally. The rise in consumer passion and purchasing power across different social classes diminishes luxury's exclusivity, leading to overconsumption and perhaps undermining potential compatibility between luxury and sustainability (Athwal et al., 2019). The luxury sector received growing public critiques on unsustainable practices in recent years, such as animal abuse, labour exploitation, and lack of supply chain transparency (Cheah et al., 2016).

The literature has confirmed the significance of the luxury industry by recognising its trickle-down effect on mainstream markets in terms of innovations, circular economy business models, and sustainable consumer behaviours (Athwal et al., 2019; Osburg et al., 2021). Consumers, particularly Generation Z, are becoming environmentally and ethically conscious and willing to buy luxury goods labelled as eco-friendly or sustainable (Bain and Company, 2021; Han et al., 2017). However, despite the increasing efforts of luxury brands to promote sustainability, a gap exists between consumers' attitudes and their actual purchase behaviours regarding sustainability (Morewedge et al., 2021). Although consumers show concerns about sustainability and expect luxury fashion companies to demonstrate a social commitment, they do not exhibit sustainable behaviours when consuming luxury fashion products, representing a significant challenge in sustainable luxury marketing (Jung et al., 2021). The literature thus encourages more research on facilitating sustainable luxury consumption (Han et al., 2017). Much of research concerning the use of digital technologies in luxury marketing has so far remained as interests in improving consumer experience rather than enhancing sustainable consumption. Therefore, this chapter put forwards that VRF could be utilised to encourage SLFC and provides a theoretical basis to illustrate the ideas from which future research can be developed to answer the question:

How can VRF used in luxury fashion marketing practices contribute to promoting SLFC by stimulating consumer learning process?

In doing so, this chapter expands the scope of sustainable luxury research by assuming that VRF embedded in luxury fashion marketing activities can serve as external cues and stimuli to promote consumer learning for SLFC. To move the field forward, this chapter builds a conceptual framework that identifies prominent opportunities for future investigations to extend the theoretical perspective of consumer learning within the contextualisation paradigm of using VRF to facilitate consumer SLMC.

13.2 Theoretical Underpinnings

13.2.1 Virtual Representation of Fashion

From an ontological perspective, 'virtual' is regarded as 'not real' or 'possible', as opposed to the 'actual' physical reality which retains the properties of being 'real' (Shields, 2006). However, 'virtual' can also relate to 'ideal' or 'ideal-real' as it might be imagined beyond physical reality (Denegri-Knott & Molesworth, 2010). Therefore, informed by this ontological view of 'virtual', virtual representation in marketing could be displaying a product or service by transforming it from the physical reality into the virtual world to constitute an 'in-between' virtual and physical reality, or only in the virtual world. Such representation is supported by digital technologies that challenge conventional understandings about space, consumption, and marketing practices. These technologies can probe and perforate the boundaries between the physical and the virtual, production and consumption, object and image, enabling consumers to be less constrained by the boundaries and laws of consumption of material goods (Muhammad et al., 2021; Denegri-Knott & Molesworth, 2010). Through adopting digital technologies, a sense of telepresence could be evoked to facilitate reproduction of memories and activation of imagination (Kim & Biocca, 1997), which could be drivers of hedonic consumption experience (Hirschman & Holbrook, 1982). These technologies also encourage innovations in the concepts and beliefs about mind, body, self, and identity shown in the virtual world (Han et al., 2017).

In this chapter, fashion is learnt from the perspective of 'fashion business' rather than in the broader sociological context. VRF is conceptualised as using technologies and platforms to digitise materials and non-materials (e.g. concepts, images, ideas, goods, and services) of fashion brands to build and present counterparts in a virtual space or a hybrid physical–virtual space. It concerns the visual or intangible rendering of fashion brands and thus reanimates, recasts, and reformulates the creation, innovation, knowledge, and expertise concerning fashion in the virtual scenario. Table 13.1 shows how VRF is adopted in luxury fashion marketing touchpoints for the spatial organisation and reformulation of fashion. It also demonstrates sustainable potentials of such use of VRF.

Table 13.1 VRF stimuli utilised in luxury fashion marketing touchpoints

Type of marketing touchpoints	Purpose for involving VRF stimuli in marketing touchpoints	Technology used for VRF	Sustainable potentials of VRF
Luxury company-controlled touchpoints	Online product visualisation and product customisation service (Altarteer et al., 2016; Altarteer & Charissis, 2019)	3D VR system	⁂ Reduce the need for large stocks of different product variations ⁂ Make the customisation a timely, efficient, and enjoyable process either within a shop or from home
	Virtual fitting rooms/try-on service (Biron, 2020; Lazazzera, 2020)	VR/AR and AI	⁂ Reduce carbon emissions caused by consumers travelling to physical stores.
	Virtual products (Roberts-Islam, 2020; Hackl, 2020). ⁂ Digital-only products ⁂ Adding virtual products into existing online game environments ⁂ Offer both digital and physical versions of products	2D/3D modelling, visual technology, and AR	⁂ Decrease consumer desire for material goods ⁂ Create or change digital persona that is linked with self-concept in the physical reality (e.g. a digital persona with a sustainability mind-set) ⁂ Eliminate product disposal issues ⁂ Avoid unethical supply chains as the raw materials are coded in computers ⁂ Avoid resources needed for transportation and packaging
	Digital fashion week and Virtual fashion shows (Bailey, 2021; Browchuk, 2020)	Live streaming	⁂ Wider possibility of audiences worldwide with reduced travelling ⁂ Cost-saving ⁂ Predict precise production volumes to avoid unnecessary mass production
	Virtual stores (Bailey, 2021) In-store brand VR experiences (Morillo et al., 2019)	VR VR/AR	⁂ Reduce travelling-related carbon emissions. ⁂ Immersive and experience to educate sustainable behaviour

(continued)

Table 13.1 (continued)

Type of marketing touchpoints	Purpose for involving VRF stimuli in marketing touchpoints	Technology used for VRF	Sustainable potentials of VRF
Luxury company-uncontrolled touchpoints	Consumers' use and experience with virtual products in games or social media (Muret, 2020) Consumers' electronic word-of-mouth (eWOM) related to brands' use of VRF (Muret, 2020; Thoumrungroje, 2014)	2D/3D modelling, VR/AR, online games, social media VR/AR and social media	• Decrease demand for material goods • Sustainable habits formulation • Interpersonal influence on sustainable consumption
Company partially controlled touchpoints	Advertising or social media messages collaborated with virtual models, ambassadors, or influencers (Arsenyan & Mirowska, 2021; Hackl, 2021)	VR/AR, computer-generated imagery (CGI), and AI	• Encourage and persuade consumers' sustainable purchase and consumption

VRF can be adopted in three types of luxury fashion marketing touchpoints where consumers could interact with a brand: company-controlled, company-uncontrolled, and company partially controlled touchpoints (Baxendale et al., 2015). Company-controlled touchpoints can incorporate VRF in marketing dimensions of production, distribution, and promotion within owned media. For example, The Fabricant, a digital fashion design and animation company, provides wearable technology and garment goods that only exist in the virtual world (Hackl, 2020). VRF is here adopted in both production and consumption to build relationships among body image, science, technology, and aesthetic experience of fashion. This use of VRF may help to reduce consumers' desire for material fashion goods and avoid product disposal issues. It can also project consumers' digital persona with sustainable mind-sets. Furthermore, VRF can implement diverse brand events like virtual fashion shows (Browchuk, 2020). These events can be creative, aesthetic, critical, passionate, and provocative, providing immersive and interactive experiences to consumers by using a layered set of sensory signifiers that capture what a brand is and what it symbolises to consumers (Jung et al., 2021). VRF thus provides opportunities for consumer-brand co-creation of value and desire. The sustainable potential of such use of VRF could be cost-saving, reducing consumer travelling and subsequent decreased carbon emissions, and precisely predicting production volumes to avoid overproduction.

Company-uncontrolled touchpoints mainly concern consumer-dominated interaction with brands, such as consumption and user experience of virtual products and consumer electronic word-of-mouth (eWOM) that concern brands' use of VRF. The third type of touchpoints might be partially controlled and monitored by companies and produced by third-party stakeholders of virtual models, influencers, and ambassadors. The adoption of VRF in these uncontrolled and partially controlled types of touchpoints is expected to promote sustainable consumption indirectly through interpersonal influence and social norms. In summary, VRF can act as a means of structuring, reconfiguring, and disseminating creativity, information, knowledge, and properties of luxury fashion brands, as well as creating highly accessible and immersive sensory worlds.

13.2.2 Sustainable Luxury

'Luxury' is derived from the Latin word 'luxuria' that portrays extravagant or excessive living (Danziger, 2005). Luxurious objects are characterised as those unessential but desirable things that provide great gratification, indulgence, and sumptuousness for people's lifestyles (Li et al., 2012). However, sustainability is linked with altruism, moderation, and ethics (Achabou & Dekhili, 2013). Despite the conflicts between the attached meanings to the two concepts, sustainable luxury is a nascent area that critically considers how the scope of sustainability can be practical and broadened in the luxury sector (Athwal et al., 2019; Osburg et al., 2021). Sustainability can serve as a facilitator of innovations and developing additional value into luxury brands. Research has identified four categories of consumer perceived values of luxury brands: utilitarian, hedonic, symbolic, and economic values (Choo et al., 2012). With embedding sustainability into luxury brands, three types of additional values are recognised, including socio-cultural values regarding conspicuousness, national identity, and sense of belonging, eco-centred values concerning doing good and not doing harm, and ego-centred values related to hedonism, health and youthfulness, guilt-free pleasures, and durable quality (Hennigs et al., 2013). These values underpin the compatibility between sustainability and luxury and the bringing together these two concepts in marketing practices.

This chapter adopts the definition of sustainable luxury given by Athwal et al. (2019) that 'entails the scope of design, production and consumption that is environmentally or ethically conscious (or both) and is oriented toward correcting various perceived wrongs within the luxury industry, including animal cruelty, environmental damage and human exploitation'. Current research on sustainable luxury is rooted in different disciplines, such as design, production, engineering, supply chain management, marketing, and business ethics (Kunz et al., 2020). This chapter takes the marketing perspective to propose a framework to explain how to embed VRF stimuli in luxury fashion marketing practices to evoke and promote consumer SLFC.

13.2.3 Sustainable Luxury Fashion Consumption

The literature has recognised both opportunities and tensions within the development of consumer SLFC. Sustainable consumption supports reducing adverse environmental impacts, purchasing products with sustainable sourcing, production, and features, and also utilising more sustainable modes of product disposal (White et al., 2019). Beyond looking for the immediate benefits relevant to self, consumers' motivation for purchasing sustainable goods involves longer-term benefits to society, the environment, natural resources, and future generations. Consumer SLFC can be encouraged by a variety of factors, such as boycotting fur enforced by politics, 'buy less, buy better' facilitated by green consumerism, elitist lifestyle, and minimalism, guilt-free consumption legitimised by philanthropic actions of luxury organisations, and refinement of consumption choices promoted by access to recycled items, 'pre-loved' luxury, 'sharing economy', vintage, and rental services (Kessous & Valette-Florence, 2019; Pantano & Stylos, 2020; Turunen et al., 2020).

However, there are also issues with luxury fashion consumers choosing sustainable products. The first issue can be negative consumer perception towards the quality of sustainable luxury goods. Research found that sustainable luxury products using recycled elements or organic raw materials are perceived by consumers as of poorer quality than unsustainable luxury products, highlighting consumer perceived problematic fit between luxury and sustainability (Achabou & Dekhili, 2013). Moreover, the incompatibility between hedonism and sustainable luxury could lead to consumers being against sustainable products. Beckham and Voyer (2014) found that consumers regard these sustainable luxury goods as less desirable and luxurious. Cervellon and Shammas (2013) noted that a reduced pleasure exists when consumers use sustainable luxury products. Thus, sustainability could undermine consumer perceptions of luxuriousness, prestige, self-pleasure, and satisfaction. Consumers might prefer an unsustainable luxury product that signals higher status, power, and prestige to a sustainable one (Amatulli et al., 2020). Another suggested issue regarding consumers engaging in SLFC is viewing sustainable factors as less significant in their purchase criteria (Davies et al., 2012). This thought could be triggered by a lack of information or knowledge about sustainable luxury. Given a luxury product without an explicit indication of being sustainable, consumers may have resource-acquisition fatigue as unwilling to spend the time and money needed to explore its sustainability (Moraes et al., 2017). In addition, consumers who lack knowledge about sustainable luxury may see luxury goods as being less harmful to the environment than non-luxury goods (Vigneron & Johnson, 2004).

Overall, the literature shows that consumer attitudes and actual purchase behaviours regarding sustainable luxury brands are incompatible. Their attitudes are ethically and morally complex around sustainable concerns, whereas their behaviours may not operate in a sustainable manner. It has been suggested that sustainable consumption could be heavily influenced by context and social conventions (Han et al., 2017). A supportive context and incentives should be provided by luxury businesses, governments, and trade organisations to encourage consumers to make choices on and alter behaviours towards sustainable consumption (Morewedge et al., 2021). Although prior research has shed light on different formats of sustainable luxury consumption, the literature still calls for studies on how marketing relates to such behaviour. Therefore, this chapter responds to the calls and serves as a springboard for further research and knowledge development in sustainable luxury marketing whilst simultaneously highlighting the importance of VRF in marketing for promoting sustainable consumption.

13.2.4 Consumer Learning

A theoretical perspective of consumer learning (Hoch & Deighton, 1989; Bosangit & Demangeot, 2016) is adopted to examine the use of VRF to promote consumer SLFC. Consumer learning is about how consumers acquire the purchase and consumption knowledge or experience and utilise this knowledge/experience to inform future buying patterns and consumption behaviours (Li et al., 2003). The marketing literature confirms the necessity of promoting consumer learning as it can be a critical mediator of decision making and facilitator of companies' long-term profit performances (White et al., 2019). There are three learning theories, associative, cognitive, and pragmatic learning.

Associative learning is characterised as a relatively non-cognitive learning process that makes automatic and predictive associations among environmental cues and behavioural responses based on concepts such as similarity, contiguity, recognition, repetition, and reinforcement (Jayanti & Singh, 2010). The associative memory is accumulated with time and across experiences. It provides implications for reflexive links between individual goals to behavioural responses. For example, consumers always make associative linkages between product cues (e.g. brand names or product attributes) and consumption benefits when evaluating products. This learning theory has been adopted in consumer research to investigate underlying mechanisms of consumers' brand evaluation and choice affected by visual stimuli (Van Osselaer & Janiszewski, 2001).

Cognitive learning leads to reasoned behavioural responses and thus represents contrasting modes of learning to associative learning (Witt, 2001). It refers to a cognitively mediated learning process described as reflective, analytical, conscious, deliberate, and systematic (Jayanti & Singh, 2010). This learning is goal motivated and developed based on categorisation, memory, explanation, and analytical representation. Cognitive learning supports the exploration of consumer information-processing, consumer expertise, and inference making (Lee & Olshavsky, 1994).

Pragmatic learning concerns learning from social interactions, negotiation, and collaboration and intersects two alternative learning approaches (Jayanti & Singh, 2010). The first learning process is an acquisitive approach that focuses on the individual level and recognises learning as acquiring knowledge and skills. The second learning process is a participative approach at a collective level and considers learning as participation in communities of practice. Therefore, pragmatic learning simultaneously concentrates on how learning occurs (participative approach) and what is learned (acquisitive approach). This learning process is highly relevant to consumer community research as communities enable consumers to engage in collective productive inquiry to guide actions (Jayanti & Singh, 2010; Steils & Hanine, 2016).

Previous research has widely applied consumer learning to explore the process by which consumer attitudes and behaviours being changed by learning from multiple sources, such as direct interaction with the product/service or information and experience obtained from different marketing touchpoints including eWOM, brochures, and advertising (Van Osselaer & Alba, 2000; Iyengar et al., 2007). More important, it is evidenced that consumer learning plays a vital role for the diffusion of sustainable consumption patterns and reducing the attitude–behaviour gap in sustainable consumption (Buenstorf & Cordes, 2008; Barth et al., 2012, 2014; Kaman, 2014), providing justification for using consumer learning as the theoretical perspective in this chapter. Promoting consumer adoption of sustainable consumption and addressing unsustainable consumer behaviour remain critical challenges for sustainable luxury development. Educational efforts and a supportive context are necessitated to deal with these challenges to promote consumer SLFC behaviour. As such, literature calls for an investigation of educating sustainable consumption that facilitates consumer learners to acquire the knowledge and skills necessary for sustainable behaviours and lifestyles (Frank & Stanszus, 2019). In response, this chapter intends to build a conceptual framework to provide ideas on adopting VRF in luxury brand marketing activities to encourage consumer learning to raise their behavioural intention of SLFC. This chapter proposes that given the use of VRF in various marketing touchpoints (e.g. virtual products, virtual shops, and promotional messages), consumers can be encouraged to learn about abstract concepts regarding sustainable luxury and increase intention to buy sustainable luxury fashion products.

13.3 Conceptual Framework

Within the perspective of consumer learning, a conceptual framework (Fig. 13.1) is developed to illustrate the implication and mechanisms of VRF for promoting consumer SLFC. The abstract ideas and concepts about sustainable luxury, such as reducing the use of resources, decreasing demand for material goods, and avoiding overproduction, should be predefined and linked with various VRF stimuli (see Table 13.1). Then these VRF stimuli can be involved in the aforementioned three types of marketing touchpoints to evoke different consumer learning approaches, which are expected to lead to consumers' behavioural responses concerning SLFC. As such, further research can test whether the three learning patterns could serve as mechanisms underlying the effects of VRF stimuli

13 The Rise of Virtual Representation of Fashion in Marketing...

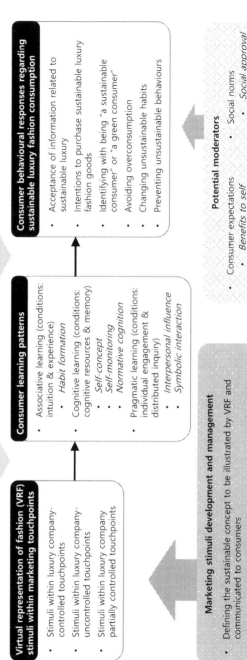

Fig. 13.1 Implications of virtual representation of fashion (VRF) for encouraging sustainable luxury fashion consumption (SLFC)

on consumer SLFC. Besides, there are factors that may influence the effectiveness of VRF stimuli and potential moderators for consumer SLFC, which can also be examined later. The following paragraphs explain how to encourage consumer SLFC as behavioural responses of three learning processes in detail.

Firstly, associative learning leads to consumers making behavioural responses based on their unconscious intuitions and experiences. Habit formation could encourage consumers who adopt this learning approach to build sustainable consumption (White et al., 2019). As relatively automatic and persistent behaviours, habits are significant for the development of SLFC as consumers making sustainable choices depend significantly more on habits than on knowledge of sustainable marketing practices (Goworek et al., 2012). Marketers can encourage consumers to engage in sustainable consumption through help consumers build positive habits and change unsustainable habits, such as using extrinsic incentives to increase desired sustainable behaviours and building regularly encountered contextual cues (e.g. provide free customisation service or rewards points to encourage consumer use of virtual try-on) to encourage action repetition.

Secondly, cognitive learning facilitates consumers' cognitive information-processing and memory, which may, in turn, encourage their reasoned behavioural responses. Within this learning process, consumers' SLFC can be facilitated by their cognitive analysis of the use of VRF in luxury company-controlled marketing touchpoints and positive connections of SLFC with their self-concepts, self-monitoring, and normative cognition. Self-concept is about a collection of beliefs, attitudes, and opinions that an individual holds concerning the cognitive perception and evaluation of self-existence (Rosenberg, 1979). Consumers may purchase and use the products that are congruent with their self-images to express parts of their self-concept or the products that could enhance their self-images to extend their self-concept (Phau & Lo, 2004). For example, as possessions, virtual luxury products can offer pathways for representation or extensions of consumers' self-concept. Moreover, the use of VRF in luxury production allows consumers to create a digital persona within a digital world. Thus, consumers can select the virtual luxury fashion product that fits or extends their self-concept and obtain ideal values that could not be maintained in physical reality.

Self-monitoring is defined by Snyder (1979) as individuals' tendency to pay close attention to appropriate behaviour in a social situation so that they can modify their behaviour to fit that situation. It underlines sensitivity to social cues as indicators of socially appropriate behaviours and the consistency level of individuals' self-presentation behaviours in different situations (Ellison et al., 2006). Consumers who embrace socially responsible luxury fashion brands may keep constant, sustainable behaviours to maintain consistency in self-concept. Moreover, consumers who are sensitive to social cues might greatly emphasise the public-self and social identities and be more willing to focus on societal welfare rather than the self-related benefits of adopting SLFC.

Normative cognition refers to the tendency of individuals to conform to other people or a reference group to gain acceptance and fulfil their expectations and differentiate themselves from others who are not highly esteemed (Vigneron & Johnson, 1999; Wilcox et al., 2009). Consumers might want to be connected with a reference group composed of people whose emotions, opinions, values, and behaviours are adopted by them for inference (Mourali et al., 2005). Accordingly, SLFC behaviours adopted by the reference group can be accepted and imitated by consumers to enhance their self-concept and obtain the reference group's approval. Therefore, using VRF stimuli in company-uncontrolled and partially controlled touchpoints might be effective here to encourage consumers' acceptance and adoption of reference groups' SLFC behaviours.

Thirdly, consumer SLFC might be promoted by interpersonal influence and symbolic interaction through their pragmatic learning process. Interpersonal influence is recognised as an indispensable scope to comprehensively understand factors affecting consumers' values, attitudes, and behaviours (Mourali et al., 2005). As a form of social influence, it refers to the tendency to perceive information obtained from other individuals as accurate statements of reality (Deutsch & Gerard, 1955). It can take many forms, such as conformity (Jahoda, 1959), peer pressure (Estrada & Vargas-Estrada, 2013), followership and leadership (Oc & Bashshur, 2013), and persuasion (Cascio et al., 2015). Consumers frequently refer to others' evaluations of a product/brand as a source of information to infer the quality and value of this product/brand. For example, consumers may seek information about sustainable luxury from professionals and knowledgeable others to make judgements. Through their pragmatic learning, consumers associate newly acquired and previously known information and then integrate them by finding connections and conflicts. Changes may occur in consumers' feelings, thoughts, attitudes, and behaviours if the consumers perceive the information obtained from others as being inherently contributory to the achievement of his/her goals, such as improving knowledge or coping with problems (Wheeler, 2009). Moreover, symbolic interaction might be useful to evoke consumer SLFC. Symbols are culturally derived objects about which commonly shared meanings are generated, used, and maintained through individuals' social interaction (Kaiser et al., 1991). For example, sustainable luxury fashion brands could be adopted to signal consumers' sustainable identities within online social interactions. By constantly acting and interacting, social norms and values about sustainability are created, and consumer SLFC are encouraged.

13.4 Conclusion

As growing attention is given to luxury fashion brands' sustainability, marketers need strategies to communicate their sustainable commitments effectively to consumers whilst promoting sustainable consumption. This chapter is motivated by the need for a comprehensive review and framework related to the key drivers and contributors of consumer SLFC. This chapter adopts the perspective of consumer learning to propose that the use of VRF in luxury marketing touchpoints could raise consumers SLFC through encouraging consumer learning. A conceptual framework is created to illustrate directions for further academic investigation and provide implications for practical development on encouraging sustainable consumption within the luxury fashion sector.

This chapter also provides practical implications for encouraging sustainable consumption patterns for luxury fashion brands. Digital technologies could be used for VRF in various marketing dimensions, such as production, distribution, and promotion, to evoke consumer learning processes towards sustainable consumption. In addition, consumer learning could be strengthened by marketing efforts on formatting positive habits about SLFC, self-concept maintenance and development, and reference groups' interpersonal influence. Besides, a supportive context that encourages consumer learning about SLFC should be provided, and the use of VRF stimuli in different marketing touchpoints should be carefully developed, monitored, and managed.

References

Achabou, M. A., & Dekhili, S. (2013). Luxury and sustainable development: Is there a match? *Journal of Business Research, 66*(10), 1896–1903.

Altarteer, S., & Charissis, V. (2019). Technology acceptance model for 3D virtual reality system in luxury brands online stores. *IEE Access, 7*, 64053–64062.

Altarteer, S., Vassilis, C., Harrison, D., & Chan, W. (2016). Product customisation: Virtual reality and new opportunities for luxury brands online trading. In *Proceedings of the 21st International Conference on Web3D technology* (pp. 173–174). Springer.

Amatulli, C., De Angelis, M., Pino, G., & Guido, G. (2020). An investigation of unsustainable luxury: How guilt drives negative word-of-mouth. *International Journal of Research in Marketing, 37*(4), 821–836.

Ameen, N., Hosany, S., & Tarhini, A. (2021). Consumer interaction with cutting-edge technologies: Implications for future research. *Computers in Human Behavior, 120*(106), 761.

Arsenyan, J., & Mirowska, A. (2021). Almost human? A comparative case study on the social media presence of virtual influencers. *International Journal of Human-Computer Studies, 155*(102), 694.

Athwal, N., Wells, V. K., Carrigan, M., & Henninger, C. E. (2019). Sustainable luxury marketing: A synthesis and research agenda. *International Journal of Management Reviews, 21*(4), 405–426.

Bailey, C. (2021). *How emperia uses VR to market art and luxury fashion brands during COVID.* [online] AIXR. Retrieved from https://aixr.org/insights/how-emperia-uses-vr-to-market-art-and-luxury-fashion-brands-during-covid/

Bain and Company. (2021). *LuxCo 2030: A vision of sustainable luxury.* [online] Bain and Company. Retrieved from https://www.bain.com/globalassets/noindex/2021/bain_brief_luxco_2030_a_vision_of-sustainable-luxury.pdf

Barth, M., Fischer, D., Michelsen, G., Nemnich, C., & Rode, H. (2012). Tackling the knowledge–action gap in sustainable consumption: insights from a participatory school programme'. *Journal of Education for Sustainable Development, 6*(2), 301–312.

Barth, M., Adomßent, M., Fischer, D., Richter, S., & Rieckmann, M. (2014). Learning to change universities from within: A service-learning perspective on promoting sustainable consumption in higher education. *Journal of Cleaner Production, 62*, 72–81.

Baxendale, S., Macdonald, E. K., & Wilson, H. N. (2015). The impact of different touchpoints on brand consideration. *Journal of Retailing, 91*(2), 235–253.

Bazi, S., Filieri, R., & Gorton, M. (2020). Customers' motivation to engage with luxury brands on social media. *Journal of Business Research, 112*, 223–235.

Beckham, D., & Voyer, B. G. (2014). Can sustainability be luxurious? A mixed-method investigation of implicit and explicit attitudes towards sustainable luxury consumption. *Advances in Consumer Research, 42*, 245–250.

Biron, B. (2020). *Retailers like Macy's, Adidas, and Modcloth are turning to virtual fitting rooms to let consumers "try on" clothing before buying it online.* [online] Business Insider. Retrieved from https://www.businessinsider.com/retailers-like-macys-adidas-are-turning-to-virtual-fitting-rooms-2020-8

Bosangit, C., & Demangeot, C. (2016). Exploring reflective learning during the extended consumption of life experiences. *Journal of Business Research, 69*(1), 208–215.

Browchuk, E. (2020). *Digital fashion weeks and virtual shows: A rundown of fashion's new rhythm.* [online] Vogue. Retrieved from https://www.vogue.com/article/digital-fashion-weeks-2020

Buenstorf, G., & Cordes, C. (2008). Can sustainable consumption be learned? A model of cultural evolution. *Ecological Economics, 67*(4), 646–657.

Cascio, C. N., Scholz, C., & Falk, E. B. (2015). Social influence and the brain: Persuasion, susceptibility to influence and retransmission. *Current Opinion in Behavioral Sciences, 3*, 51–57.

Cervellon, M. C., & Shammas, L. (2013). The value of sustainable luxury in mature markets: A customer-based approach. *Journal of Corporate Citizenship, 52*(12), 90–101.

Cheah, I., Zainol, Z., & Phau, I. (2016). Conceptualizing country-of-ingredient authenticity of luxury brands. *Journal of Business Research, 69*(12), 5819–5826.

Choo, H. J., Moon, H., Kim, H., & Yoon, N. (2012). Luxury customer value. *Journal of Fashion Marketing and Management, 16*(1), 81–101.

Danziger, P. N. (2005). *Let them eat cake: Marketing luxury to the masses–As well as the classes.* Dearborn Trading.

Davies, I. A., Lee, Z., & Ahonkhai, I. (2012). Do consumers care about ethical-luxury? *Journal of Business Ethics, 106*(1), 37–51.

Deloitte. (2020). *Global powers of luxury goods 2020*. [online] Deloitte. Retrieved from https://www2.deloitte.com/content/dam/Deloitte/at/Documents/consumer-business/at-global-powers-luxury-goods-2020.pdf

Denegri-Knott, J., & Molesworth, M. (2010). Concepts and practices of digital virtual consumption. Consumption. *Markets and Culture, 13*(2), 109–132.

Deutsch, M., & Gerard, H. B. (1955). A study of normative and informational social influences upon individual judgment. *The Journal of Abnormal and Social Psychology, 51*(3), 629–636.

Ellison, N., Heino, R., & Gibbs, J. (2006). Managing impressions online: Self-presentation processes in the online dating environment. *Journal of Computer-Mediated Communication, 11*(2), 415–441.

Estrada, E., & Vargas-Estrada, E. (2013). How peer pressure shapes consensus, leadership and innovations in social groups. *Scientific Reports, 3*(1), 1–6.

Frank, P., & Stanszus, L. S. (2019). Transforming consumer behavior: Introducing self-inquiry-based and self-experience-based learning for building personal competencies for sustainable consumption. *Sustainability, 11*(9), 1–19.

Goworek, H., Fisher, T., Cooper, T., Woodward, S., & Hiller, A. (2012). The sustainable clothing market: an evaluation of potential strategies for UK retailers. *International Journal of Retail & Distribution Management, 40*(12), 935–955.

Hackl, C. (2020). *Why virtual dresses and augmented fashion are a new profitable frontier for brands*. [online]. Forbes. Retrieved October 28, 2021, from https://www.forbes.com/sites/cathyhackl/2020/06/08/why-virtual-dresses%2D%2Daugmented-fashion-are-a-new-profitable-frontier-for-brands/?sh=38815f2862c8

Hackl, C. (2021). *How brands can thrive in the direct to avatar economy*. [online] Forbes. Retrieved from https://www.forbes.com/sites/cathyhackl/2021/01/29/how-brands-can-thrive-in-the-direct-to-avatar-economy/?sh=3d7fe8c5417c

Han, J., Seo, Y., & Ko, E. (2017). Staging luxury experiences for understanding sustainable fashion consumption: A balance theory application. *Journal of Business Research, 74*, 162–167.

Hennigs, N., Wiedmann, K. P., Klarmann, C., & Behrens, S. (2013). Sustainability as part of the luxury essence: Delivering value through social and environmental excellence. *Journal of Corporate Citizenship, 52*, 25–35.

Hirschman, E. C., & Holbrook, M. B. (1982). Hedonic consumption: emerging concepts, methods and propositions. *Journal of Marketing*, 92–101.

Hoch, S. J., & Deighton, J. (1989). Managing what consumers learn from experience. *Journal of Marketing, 53*(2), 1–20.

Holmqvist, J., Wirtz, J., & Fritze, M. P. (2020). Luxury in the digital age: A multi-actor service encounter perspective. *Journal of Business Research, 121*, 747–756.

Iyengar, R., Ansari, A., & Gupta, S. (2007). A model of consumer learning for service quality and usage. *Journal of Marketing Research, 44*(4), 529–544.

Jahoda, M. (1959). Conformity and independence: A psychological analysis. *Human Relations, 12*(2), 99–120.

Javornik, A., Duffy, K., Rokka, J., Scholz, J., Nobbs, K., Motala, A., & Goldenberg, A. (2021). Strategic approaches to augmented reality deployment by luxury brands. *Journal of Business Research, 136*, 284–292.

Jayanti, R. K., & Singh, J. (2010). Pragmatic learning theory: An inquiry-action framework for distributed consumer learning in online communities. *Journal of Consumer Research, 36*(6), 1058–1081.

Jung, J., Yu, J., Seo, Y., & Ko, E. (2021). Consumer experiences of virtual reality: Insights from VR luxury brand fashion shows. *Journal of Business Research, 130*, 517–524.

Kaiser, S. B., Nagasawa, R. H., & Hutton, S. S. (1991). Fashion, postmodernity and personal appearance: A symbolic interactionist formulation. *Symbolic Interaction, 14*(2), 165–185.

Kaman, L. (2014). Predictors of sustainable consumption among young educated consumers in Hong Kong. *Journal of International Consumer Marketing, 26*(3), 217–238.

Kessous, A., & Valette-Florence, P. (2019). "From Prada to Nada": Consumers and their luxury products: A contrast between second-hand and first-hand luxury products. *Journal of Business Research, 102*, 313–327.

Kim, T., & Biocca, F. (1997). Telepresence via television: Two dimensions of telepresence may have different connections to memory and persuasion. *Journal of Computer-Mediated Communication, 3*(2), JCMC325.

Kunz, J., May, S., & Schmidt, H. J. (2020). Sustainable luxury: current status and perspectives for future research. *Business Research, 13*(2), 541–601.

Lazazzera, M. (2020). *How virtual stores became a reality in the world of luxury.* [online] www.ft.com. Retrieved from https://www.ft.com/content/ca6bb85f-9af7-4df7-a606-828ceeea5a97

Lee, D. H., & Olshavsky, R. W. (1994). Toward a predictive model of the consumer inference process: The role of expertise. *Psychology & Marketing, 11*(2), 109–127.

Li, H., Daugherty, T., & Biocca, F. (2003). The role of virtual experience in consumer learning. *Journal of Consumer Psychology, 13*(4), 395–407.
Li, G., Li, G., & Kambele, Z. (2012). Luxury fashion brand consumers in China: Perceived value, fashion lifestyle, and willingness to pay. *Journal of Business Research, 65*(10), 1516–1522.
Moraes, C., Carrigan, M., Bosangit, C., Ferreira, C., & McGrath, M. (2017). Understanding ethical luxury consumption through practice theories: A study of fine jewellery purchases. *Journal of Business Ethics, 145*(3), 525–543.
Morewedge, C. K., Monga, A., Palmatier, R. W., Shu, S. B., & Small, D. A. (2021). Evolution of consumption: A psychological ownership framework. *Journal of Marketing, 85*(1), 196–218.
Morillo, P., Orduña, J. M., Casas, S., & Fernández, M. (2019). A comparison study of AR applications versus pseudo-holographic systems as virtual exhibitors for luxury watch retail stores. *Multimedia Systems, 25*(4), 307–321.
Mourali, M., Laroche, M., & Pons, F. (2005). Individualistic orientation and consumer susceptibility to interpersonal influence. *Journal of Services Marketing, 19*(3), 164–173.
Muhammad, S. S., Dey, B. L., Kamal, M. M., & Alwi, S. F. S. (2021). Consumer engagement with social media platforms: A study of the influence of attitudinal components on cutting edge technology adaptation behaviour. *Computers in Human Behavior, 121*(106), 802.
Muret, D. (2020). *Fashion and luxury industries show increasing interest in gaming*. [online]. Retrieved from https://ww.fashionnetwork.com/news/Fashion-and-luxury-industries-show-increasing-interest-in-gaming,1216088.html
Oc, B., & Bashshur, M. R. (2013). Followership, leadership and social influence. *The Leadership Quarterly, 24*(6), 919–934.
Osburg, V. S., Davies, I., Yoganathan, V., & McLeay, F. (2021). Perspectives, opportunities and tensions in ethical and sustainable luxury: introduction to the thematic symposium. *Journal of Business Ethics, 169*(2), 201–210.
Pantano, E., & Stylos, N. (2020). The Cinderella moment: Exploring consumers' motivations to engage with renting as collaborative luxury consumption mode. *Psychology & Marketing, 37*(5), 740–753.
Phau, I., & Lo, C. C. (2004). Profiling fashion innovators: A study of self-concept, impulse buying and Internet purchase intent. *Journal of Fashion Marketing and Management, 8*(4), 399–411.
Roberts-Islam, B. (2020). *How digital fashion could replace fast fashion, and the startup paving the way.* [online]. Forbes. Retrieved October 28, 2021, from https://www.forbes.com/sites/brookerobertsislam/2020/08/21/

how-digital-fashion-could-replace-fast-fashion-and-the-startup-paving-the-way/?sh=4751de1670d8

Rosenberg, M. (1979). *Conceiving the self*. Basic Book.

Shields, R. (2006). Boundary-thinking in theories of the present: the virtuality of reflexive modernization. *European Journal of Social Theory, 9*(2), 223–237.

Silverstein, M. J., & Fiske, N. (2003). Luxury for the masses. *Harvard Business Review, 81*(4), 48–59.

Snyder, M. (1979). Self-monitoring processes. *Advances in Experimental Social Psychology, 12*, 85–128.

Steils, N., & Hanine, S. (2016). Creative contests: knowledge generation and underlying learning dynamics for idea generation. *Journal of Marketing Management, 32*(17–18), 1647–1669.

Thoumrungroje, A. (2014). The influence of social media intensity and EWOM on conspicuous consumption. *Procedia-Social and Behavioral Sciences, 148*, 7–15.

Turunen, L. L. M., Cervellon, M. C., & Carey, L. D. (2020). Selling second-hand luxury: Empowerment and enactment of social roles. *Journal of Business Research, 116*, 474–481.

Van Osselaer, S. M., & Alba, J. W. (2000). Consumer learning and brand equity. *Journal of Consumer Research, 27*(1), 1–16.

Van Osselaer, S. M., & Janiszewski, C. (2001). Two ways of learning brand associations. *Journal of Consumer Research, 28*(2), 202–223.

Vigneron, F., & Johnson, L. W. (1999). A review and a conceptual framework of prestige-seeking consumer behavior. *Academy of Marketing Science Review, 1*(1), 1–15.

Vigneron, F., & Johnson, L. W. (2004). Measuring perceptions of brand luxury. *Journal of Brand Management, 11*(6), 484–506.

Wheeler, L. (2009). Interpersonal influence. In H. T. Reis & S. Sprecher (Eds.), *Encyclopedia of human relationships*. Sage.

White, K., Habib, R., & Hardisty, D. J. (2019). How to SHIFT consumer behaviors to be more sustainable: A literature review and guiding framework. *Journal of Marketing, 83*(3), 22–49.

Wilcox, K., Kim, H. M., & Sen, S. (2009). Why do consumers buy counterfeit luxury brands? *Journal of Marketing Research, 46*(2), 247–259.

Witt, U. (2001). Learning to consume: A theory of wants and the growth of demand. *Journal of Evolutionary Economics, 11*(1), 23–36.

Zhang, L., & Zhao, H. (2019). Personal value vs. luxury value: What are Chinese luxury consumers shopping for when buying luxury fashion goods? *Journal of Retailing and Consumer Services, 51*, 62–71.

Index[1]

A

Activism, 220, 224–234

B

Beauty, 7, 8, 64, 69, 70, 156, 173–192
Behavioural patterns, 8, 190–192
Belief, 53, 62, 66, 89, 92, 93, 130, 132, 135, 144, 221, 261, 264, 274, 284
Blood diamonds, 154, 167
Brand equity, 63, 68, 271, 272
Branding, 223, 225
Business, 4, 16, 17, 21, 23, 26–28, 35, 36, 46, 53, 60, 85, 102, 107–109, 111, 113–116, 118, 121, 133, 134, 161, 166, 174, 176, 199, 220–223, 225–230, 233, 245, 274, 278, 280
Business models, 15, 17, 20, 23, 28, 86, 106, 115, 233, 243, 273

C

Canadian ethical diamonds, 153–169
China, 83, 141, 242, 244, 245, 247, 251, 261, 264
Circular economy, 14, 101–111, 113–121, 231, 273
Consumer behaviour, 4, 5, 88, 246, 247, 273, 282
Consumer learning, 273, 280–282, 286, 287
Consumer perception, 192, 202, 212, 213, 279

---

[1] Note: Page numbers followed by 'n' refer to notes.

© The Author(s), under exclusive license to Springer Nature Switzerland AG 2022
C. E. Henninger, N. K. Athwal (eds.), *Sustainable Luxury*, Palgrave Advances in Luxury, https://doi.org/10.1007/978-3-031-06928-4

293

Index

Consumer/s, 3–9, 14, 15, 17, 18, 25, 35, 36, 38–41, 43–49, 51–55, 60–63, 66–68, 71–73, 81–84, 87–90, 92, 93, 103, 104, 107–110, 115, 120, 121, 129–131, 133, 135, 136, 142–146, 153–169, 173–192, 197–213, 220–223, 225, 226, 228, 232–234, 241–264, 271–274, 277–282, 284–286

Consumption, 4, 7, 8, 14, 17, 38–40, 47, 48, 52, 60–64, 66, 67, 71, 82, 83, 86, 88–90, 92, 93, 102–104, 107, 109, 110, 130, 133, 154–156, 167, 173–180, 183–186, 188–192, 200, 202, 203, 219–234, 242–244, 246, 261, 271–287

Cosmetics industry, 177

Culture, 2, 4, 6, 65–68, 70, 87, 101, 115, 117, 130, 132–135, 143–145, 167, 168, 222, 226, 227, 242, 264

D

Decision-making process, 200, 241–264

Diamonds, 7, 68, 70, 153–169

Digitalisation, 9, 15–17

Durability, 45, 48, 64, 69, 72, 141, 272

E

Eco-label, 8, 198–203, 205–213, 206n1

Eco-labelling, 7, 8, 197–213

Eco-luxury, 35–55

Entrepreneurs, 7, 86, 101–121, 165

Ethics, 64, 109, 153, 156, 160, 162, 179, 223, 229–231, 278

F

Fashion, vii, 1–9, 14–21, 23, 24, 26–28, 38, 41, 45–48, 50–53, 61, 64, 66, 81–93, 101–111, 113–121, 130, 175–180, 188, 192, 197–213, 219–234, 241–264, 271–287

Fast moving consumer goods (FMCG), 176, 177, 185

Female, 41, 66, 101–121, 139, 140, 180

G

Ghana, 7, 130, 132, 138

Government, 60, 69, 81, 83, 85, 87, 89, 90, 92, 93, 102, 107–108, 111, 116–118, 121, 154, 156, 174, 220, 245, 280

H

Heritage, 7, 61, 64, 69, 70, 87, 129–146, 243, 254

History, 24, 64, 130, 144, 146, 154, 158, 165, 207, 208, 232, 254, 255, 262

I

Identity obsession, 153–169

Index

Industry, 1, 3–9, 14, 17, 24, 27, 28, 36, 50, 60–63, 68, 72, 81–87, 90–92, 101–111, 113–116, 118, 119, 121, 139, 153–155, 160, 166, 167, 173–192, 198, 200–203, 205, 206, 208–213, 220, 227–229, 241, 242, 244, 273, 278
International, 4, 5, 20, 40, 107, 108, 115, 154, 175, 222
Islam, 82, 93
Italy, 153–169

Jewellery, 7, 68, 70, 89, 92, 130–132, 135, 137, 143, 144, 244

Kente, 7, 129–146

Labelling, 187, 189, 191, 199
Language, 6, 51, 53, 65, 66, 68, 70, 71, 176
Lifestyle, 35–39, 43, 46–47, 49, 52, 54, 64, 103, 180, 191, 192, 203, 278, 279, 282
Little Red Book, 245, 247, 248, 256, 257, 260, 261
Luxury, vii, 1–9, 13–29, 35, 39–44, 51–54, 59–73, 81–93, 101–121, 129–146, 156, 157, 161, 173–192, 197–213, 219–234, 241–264, 271–287

Luxury beauty, 8, 173–192
Luxury brand communication, 178

Marimekko, 21, 23, 43, 49–51, 53
Marketing, 4, 9, 83, 106, 116, 154, 157, 176, 201, 208, 221, 223, 225, 259, 263, 264, 271–287
Marketing practice, 271–287
Marketing stimuli, 273, 275–278, 287
Muslim, 82, 88, 89, 92, 93

Natural dyes, 36, 50

Organisation, 85, 109, 117, 192, 198, 221, 224, 230, 231, 274, 279, 280
Oxymoron, 81–93

Paradox, 6, 82, 83, 92
Partnership, 16, 21, 24–26, 28, 62–63, 108, 201
Platform, 13–19, 21, 23–29, 41, 110, 179, 180, 224, 228, 241–245, 256, 261, 264, 272, 274
Policy, 84, 92, 102, 103, 116, 117, 119, 161, 174, 177, 225, 227, 229

Production process, 6, 7, 47, 50, 87, 109, 130–132, 134–140, 143, 144, 186
Product-service systems, 17

R
Rarity, 64, 69, 70, 250, 272
Raw materials, 4, 18, 87, 90, 106, 115, 116, 118, 119, 132–136, 138–141, 145, 201, 279
Recycling, 17, 18, 23, 66, 83–85, 102, 103, 105, 109, 111, 116–117, 119, 121, 139, 186, 189
Resale, 13–29, 110

S
Saudi Arabia, 6, 81–93, 101–121
Scandinavian, 35–55
Second-hand, 5, 8, 9, 13–29, 88, 93, 103, 105, 106, 109, 110, 117, 138, 139, 241–264
Slow design, 53
Social media, 8, 38, 41, 110, 241–264, 271, 272
Social sustainability, 8, 89, 107, 134, 136, 138, 139, 219–234
Social sustainable consumption, 221, 280
Supply chain, 3, 7, 15, 18, 19, 21, 23, 45, 46, 48, 53, 70, 109, 118, 129–146, 153, 155–157, 160, 162, 166, 186, 219, 221, 229, 273, 278
Sustainability, 1–4, 6–9, 14, 15, 17, 18, 25, 28, 29, 39, 45, 46, 48, 53, 54, 59–73, 81–93, 101–108, 110, 113–116, 118, 119, 129–146, 153, 160, 167, 173–192, 197–199, 201, 203, 205, 206, 208, 219–234, 249, 252, 261, 262, 272, 273, 278–280, 286
Sustainability claims, 68, 69, 173–192
Sustainability communication, 191, 198, 200, 209–212
Sustainability information, 88, 90, 181, 200, 202, 206
Sustainability performance, 198, 202, 211
Sustainability signals, 8, 175, 181, 183–188, 190, 192
Sustainable luxury, vii, 1–4, 6, 9, 59–73, 82, 116, 191, 202, 219, 272, 273, 278–280, 282, 285, 286
Sustainable luxury fashion consumption (SLFC), 9, 271–287

T
Textile, 4, 7, 17, 27, 36, 38, 39, 41, 45, 49, 50, 87, 106, 109, 118, 121, 129–146
Tiffany, 68–71, 154, 156
Timelessness, 64, 69, 272
Traceability, 7, 8, 153–169, 249, 254–255, 262

V
Value congruence, 59–73

Values, 5, 13–29, 35–38, 40, 41, 43, 47, 48, 50–55, 59–73, 86, 101, 102, 106, 108–110, 116, 129, 132, 135, 143–145, 155, 156, 160, 161, 165–168, 173–175, 178, 184, 186, 190–192, 202, 219, 221–223, 229–231, 243, 244, 249–252, 255, 261, 262, 271, 277, 278, 284–286

Virtual representation, 271–287
Vision 2030, 83–86, 90, 92, 101–103, 107, 111, 113–117, 119
Visual identity, 133–145

W

Weibo, 245, 247
Woke brands, 223